GLOBAL SOURCES OF LOCAL POLLUTION

An Assessment of Long-Range Transport of Key Air Pollutants to and from the United States

Committee on the Significance of International Transport of Air Pollutants

Board on Atmospheric Sciences and Climate

Division on Earth and Life Studies

NATIONAL RESEARCH COUNCIL
OF THE NATIONAL ACADEMIES

THE NATIONAL ACADEMIES PRESS
Washington, D.C.
www.nap.edu

THE NATIONAL ACADEMIES PRESS   500 Fifth Street, N.W.   Washington, DC 20001

This study was supported by the U.S. Environmental Protection Agency under contract number EP-D-08-0530. Any opinions, findings, and conclusions, or recommendations expressed in this material are those of the author(s) and do not necessarily reflect the views of the intelligence community or any of its sub-agencies.

International Standard Book Number 13: 978-0-309-14401-8
International Standard Book Number 10: 0-309-14401-9
Library of Congress Control Number: 2009941434

Additional copies of this report are available from the National Academies Press, 500 Fifth Street, N.W., Lockbox 285, Washington, DC 20055; (800) 624-6242 or (202) 334-3313 (in the Washington metropolitan area); Internet, http://www.nap.edu.

THE NATIONAL ACADEMIES
Advisers to the Nation on Science, Engineering, and Medicine

The **National Academy of Sciences** is a private, nonprofit, self-perpetuating society of distinguished scholars engaged in scientific and engineering research, dedicated to the furtherance of science and technology and to their use for the general welfare. Upon the authority of the charter granted to it by the Congress in 1863, the Academy has a mandate that requires it to advise the federal government on scientific and technical matters. Dr. Ralph J. Cicerone is president of the National Academy of Sciences.

The **National Academy of Engineering** was established in 1964, under the charter of the National Academy of Sciences, as a parallel organization of outstanding engineers. It is autonomous in its administration and in the selection of its members, sharing with the National Academy of Sciences the responsibility for advising the federal government. The National Academy of Engineering also sponsors engineering programs aimed at meeting national needs, encourages education and research, and recognizes the superior achievements of engineers. Dr. Charles M. Vest is president of the National Academy of Engineering.

The **Institute of Medicine** was established in 1970 by the National Academy of Sciences to secure the services of eminent members of appropriate professions in the examination of policy matters pertaining to the health of the public. The Institute acts under the responsibility given to the National Academy of Sciences by its congressional charter to be an adviser to the federal government and, upon its own initiative, to identify issues of medical care, research, and education. Dr. Harvey V. Fineberg is president of the Institute of Medicine.

The **National Research Council** was organized by the National Academy of Sciences in 1916 to associate the broad community of science and technology with the Academy's purposes of furthering knowledge and advising the federal government. Functioning in accordance with general policies determined by the Academy, the Council has become the principal operating agency of both the National Academy of Sciences and the National Academy of Engineering in providing services to the government, the public, and the scientific and engineering communities. The Council is administered jointly by both Academies and the Institute of Medicine. Dr. Ralph J. Cicerone and Dr. Charles M. Vest are chair and vice chair, respectively, of the National Research Council.

www.national-academies.org

vii

Acknowledgments

This report has been reviewed in draft form by individuals chosen for their diverse perspectives and technical expertise, in accordance with procedures approved by the National Research Council's Report Review Committee. The purpose of this independent review is to provide candid and critical comments that will assist the institution in making its published report as sound as possible and to ensure that the report meets institutional standards for objectivity, evidence, and responsiveness to the study charge. The review comments and draft manuscript remain confidential to protect the integrity of the deliberative process.

We wish to thank the following individuals for their review of this report:

BETSY ANCKER-JOHNSON, General Motors Corporation (retired), Austin, Texas

ELLIOT ATLAS, University of Miami, Florida

RICHARD BURNETT, Health Canada, Ottawa

ELVIN HEIBERG III, Heiberg Associates. Inc., Arlington, Virginia

RUDOLPH HUSAR, Washington University, St. Louis, Missouri

QINBIN LI, Univeristy of California, Los Angeles

ARTHUR LEE, Chevron Corporation, San Ramon, California

DENISE MAUZERALL, Princeton University, Princeton, New Jersey

NICOLA PIRRONE, CNR-Institute for Atmosperhic Pollution, Monterotondo, Italy

YANNIS YORTSOS, University of Southern California, Los Angeles

Although the reviewers listed above have provided many constructive comments and suggestions, they were not asked to endorse the conclusions or recommendations, nor did they see the final draft of the report before its release. The review of this report was overseen by **William Randel**, National Center for Atmospheric Research. Appointed by the National Research Council, he was responsible for making certain that an independent examination of this report was carried out in accordance with institutional procedures and that all review comments were carefully considered. Responsibility for the final content of this report rests entirely with the authoring committee and the institution.

We offer our sincere thanks to the staff who supported the work of this committee, including our study director, Laurie Geller, and our Program Associate, Rob Greenway. We also thank the many people who offered direct input to the committee with meeting presentations and personal, phone, or email discussions, including: Terry Keating, U.S. Environmental Protection Agency; John Bachmann, Vision Air Consulting; Dan Reifsnyder and Susan Gardener, U.S. Department of State; Bryan Wood-Thomas, American Shipping Council; Joe Prospero, University of Miami; Arlene Fiore and Hiram Levy, National Oceanic and Atmospheric Administration (NOAA)/Geophysical Fluid Dynamics Laboratory; Hongbin Yu, Drew Shindell, and Jim Crawford, National Aeronautics and Space Administration; A.R. Ravishankara, NOAA; Denise Mauzerall, Princeton University; Chris Shaver, Colleen Flanagan, and Ellen Porter, National Park Service; Kim Prather, Scripps Institution of Oceanography; and Sherwood Rowland, University of California, Irvine.

Contents

xi

Summary

Ten days ago factories, traffic, and cooking stoves half a world away emitted pollutants that may make the air you inhale today more hazardous to your health. Air pollutants emitted by your local electric power plant or your lawn mower today may help push air pollution levels above local air quality standards somewhere in Europe next week. Recent advances in air pollution monitoring methods and models, including satellite-borne sensors, have made it possible to show that long-range transport of air pollutants, including transpacific and transatlantic trajectories, does occur. These studies make it clear that the world has only one atmosphere and that adverse impacts of emitted pollutants often cannot be confined to one location, one region, or even one continent.

Many types of air pollutants have been observed to travel far from their sources. These include primary pollutants that are directly emitted like soot particles from diesel vehicles, windblown dust from deserts or degraded croplands, mercury from coal-fired power plants, pesticides from agricultural operations, and nitrogen oxides from motor vehicles. They also include secondary pollutants that are created in the atmosphere by chemical reaction sequences that begin with primary pollutants. Important secondary pollutants include atmospheric oxidants like ozone and hydrogen peroxide, sulfuric and nitric acids, and chemically diverse secondary smog particles. Any air pollutant with an atmospheric lifetime of at least three to four days may be transported across most of a continent, a week or two may get it across an ocean, a month or two can send it around the hemisphere, and a year or two may deliver it anywhere on Earth.

Both population and living standards are increasing in many parts of the world; often, so are the resulting pollutant emissions. At the same time, air quality standards are being tightened in response to studies that demonstrate adverse health consequences at ever lower exposures. Other studies document unwelcome impacts of air pollutants on crop yields and the viability of forests, grasslands, and other natural ecosystems. Still other studies are unveiling the interplay between air pollution levels and climate change on scales ranging from regional to global. All of these concerns have led to increasing international efforts to recognize and measure long-range transport of pollutants. They have also spurred attempts to predict how expected changes in population; in production of food, energy, and goods; and in climate will impact future pollutant transport and air quality.

REPORT GENESIS

Federal, state, and local agencies are faced with the need to understand and manage the current impact of long-range transport of pollution on health and well-being (relative to the impacts of local pollution sources), as well as the need to develop measurement methods and models to track pollution transport trends and project their future levels and impacts. Agencies with environmental regulatory responsibilities need to better understand how, when, and where long-range transport may lead to National Ambient Air Quality Standard (NAAQS) violations or pollutant levels that exceed other regulatory guidelines. Agencies with atmospheric research portfolios are interested in knowing what gaps in knowledge and capability weaken our understanding of current long-range transport of pollutants, as well as those that reduce our abilities to measure trends and predict future activity and its impacts. They would also like to envision how our atmospheric observational capabilities and diagnostic and predictive models can be improved to close these gaps.

In response to these challenges the Environmental Protection Agency (EPA), the National Oceanic and Atmospheric Administration (NOAA), the National Aeronautics and Space Administration (NASA), and the National Science Foundation (NSF) have cosponsored this National Research Council (NRC) Committee to explore these issues for four specific pollutant classes: ozone and it precursors (O_3), fine particulate matter and its precursors (PM), atomic and molecular mercury (Hg), and persistent organic pollutants (POPs).[1] Specifically, the Committee was asked to consider the impacts of long-range pollution transport on air quality, pollutant deposition, and radia-

---

[1] Long-lived greenhouse gases, such as CO_2 (recently ruled to be a "pollutant" by the U.S. Supreme Court), have long been known to undergo global-scale transport, but the Committee's charge did not include consideration of such gases.

tive forcing[2] in the United States; the impacts of U.S. emissions on foreign air and environmental quality; and the ways these emissions and impacts may change in the future. Finally they were asked to identify how better research and information management tools might improve understanding of and the ability to quantify the impacts and implications of long-range transport of pollution. The complete Statement of Task is presented in Appendix A.

KEY CONCLUSIONS

Both observational and modeling studies confirm that significant quantities of the pollutants considered in this study are transported over long distances, both to and from North America. Meteorological conditions off the east coasts of both Asia and North America are conducive to lofting pollutants into the midlatitude free troposphere, where strong winds aloft can rapidly transport polluted Asian air masses to North America and polluted North American air masses to Europe. Satellite observations and high-altitude in situ measurements observe polluted air masses in the free troposphere crossing both the Pacific and Atlantic ceans. The mixing of these pollutants down to ground level is not well characterized, and thus the question of how much and where long-range transport of pollution from distant emissions can impact local air quality is much harder to determine. Other mechanisms that transport pollutants over long ranges at lower altitudes and lower latitudes are also incompletely characterized. Nonetheless, atmospheric chemical transport models do predict the occurrence of ground-level pollution due to long-range transport, and low-altitude and ground-level measurements do at times clearly detect such events. The Committee thus concludes that long-range transport of pollutants from foreign sources can under some conditions have a significant impact on U.S ambient concentrations and deposition rates for each of the four pollutant classes considered and that U.S. environmental goals are affected to varying degrees by nondomestic sources of these pollutants. Similarly, long-range transport of pollutants originating in the United States can significantly affect air quality and other environmental concerns elsewhere (e.g., northern and central Europe, the Arctic). The relative importance of long-range pollutant contributions from foreign sources is likely to increase as nations institute stricter air quality standards that result in tougher emissions controls on domestic sources.

Our current ability to fully characterize long-range transport and its impacts is limited, due to a number of factors: uncertainties in foreign

---

[2] Radiative forcing (RF) of climate is a quantitative measure (usually in watts per square meter) of the instantaneous imbalance in the climate system caused by the addition of greenhouse gases or aerosols to the atmosphere.

emission source strengths, incomplete understanding of pollutant chemical and phase transformations during transport, poorly characterized mechanisms and rates of pollutant transfer between the boundary layer and free troposphere, and the fact that very few air quality research and monitoring sites are equipped to discern long-range pollutant contributions amid the normally much larger pollutant inputs from local and regional sources. Conclusions about the overall magnitude of pollution inflow and outflow must be drawn from modeling estimates, which are constrained by many uncertainties, including those listed above.

The Committee concludes that the most effective way to improve our capability to characterize current long-range pollutant impacts and to predict future trends is the development and implementation of an integrated pollutant-source attribution observation and modeling system.

Key components must include

- timely, more highly resolved and more accurate multipollutant **emission scenarios** for all significant activity sectors (including sources associated with the natural environment) in key source regions;
- frequent **satellite observations** (and associated analyses) of a range of important primary and secondary pollutants with adequate horizontal and vertical resolution in order to identify long-range transport events and to verify regional emissions estimates with inverse modeling;
- more capable **in situ monitoring** of multipollutant ground-level and vertical profile concentration measurements at judiciously chosen receptor sites with measurement capabilities designed to fingerprint pollutants from foreign sources and provide key pollutant ratios and other data necessary to establish trends and allow source attribution calculations;
- periodic intensive **field measurement campaigns** designed to track long-range transport events, better characterize chemical and physical transformations of key pollutants, and identify and quantify mechanisms that transport pollutants from the surface into the free troposphere and back; and
- improved **chemical transport models** capable of using inversion techniques (wherein pollutant emissions are derived from measured atmospheric concentrations), to better verify emission estimates, provide better estimates of source/receptor relationships, and predict trends based on informed scenarios for future emissions and changing climate conditions.

Pollutant-Specific Findings

The assessment of long-range transport of pollution involves many issues that are common to all pollutant species (e.g., the need to quantify sources, characterize atmospheric transport pathways, define baseline concentrations, integrate observational and modeling analyses), but each type of pollutant has its own unique characteristics and environmental significance. The Committee thus developed a range of findings and recommendations for each of the four main pollutant classes considered. Detailed recommendations for future research needs are presented at the end of Chapters 2-5. Below is a brief overview of key findings regarding what is known and not known about long-range transport, and resulting impacts for each of these pollutants. The list highlights but does not fully duplicate the key findings from the individual pollutant chapters.

Ozone and Precursors

• There is clear evidence that baseline levels of tropospheric O_3 have risen above pre-industrial levels in the Northern Hemisphere by 40-100 percent, and much of this increase can be directly attributed to human-caused emissions of O_3 precursor species. Baseline tropospheric O_3 abundances at many remote locations in the Northern Hemisphere have increased over the last few decades; the rates vary by location and the causes of these recent changes are not clear.

• Plumes containing high levels of O_3 and its precursors can be transported between continents and are observed downwind of major industrial regions (and large wildfires) in North America, Europe, and Asia. These plumes are observed in the free troposphere over affected regions, but less frequently at the surface due to dilution in the boundary layer.

• Multimodel studies calculate that a 20 percent reduction in ozone precursor emissions from any three of the four major industrial regions of the Northern Hemisphere will reduce surface O_3 in the fourth region by about one part per billion (ppb) on average but with large spatial and seasonal variations.

• U.S. NAAQS ozone violations are caused primarily by domestic emissions but are augmented by a changing baseline, as well as episodic nonlocal events, such as emissions from wildfires, lightning NO_x, stratospheric intrusions, and occasional plumes from distant anthropogenic sources. Most violations are only a few ppb above the standard, and thus increasing baseline O_3 can contribute to these violations.

• It is clear that distant pollution does contribute to increased concentrations of O_3 over populated regions and that such increases may have detrimental impacts on human health, agriculture, and natural ecosystems. One study estimates that the number of premature cardiopulmonary deaths that could be avoided per year in North America due to a 20 percent emission reduction in other major Northern Hemisphere industrial regions is in the hundreds. The uncertainty in this estimate is large (at least ±50 percent), reflecting uncertainties in modeling both O_3 change and health effects.

Particulate Matter and Precursors

• Observations demonstrate that PM can undergo long-range transport, primarily through episodic events associated with biomass burning plumes, dust storms, and fast transport of industrial pollution. Such episodes are most easily identifiable in remote area observations with low ambient PM concentrations. More persistent long-range transport also occurs but at relatively low concentrations that are difficult to distinguish from local and regional PM sources. There is not enough observational evidence to demonstrate clear trends in average baseline levels of surface PM on a global or hemispheric scale.

• Chemical transport models are increasingly being used to estimate long-range transport contributions to atmospheric PM concentrations. Uncertainties in model estimates result from factors such as lack of observational constraints, particularly in the free troposphere, where much of the transport occurs; poorly constrained or unknown emissions of some primary particles and the emissions and conversion of PM precursors (especially secondary organic PM); and poorly constrained wet and dry PM removal processes as reflected in the large differences in PM lifetimes among models.

• There are growing capabilities for observing total column amounts of PM, but the linkages between column amounts and surface concentrations are not well understood, which limits the ability to quantify the effect of persistent transport on surface concentrations based on column observations.

• With the exception of occasional extreme episodes, long-range transport of PM is estimated to represent a negligible contribution to PM NAAQS exceedances in most regions of the United States. However, increases in surface PM from episodic or persistent transport can be significant in places with low ambient pollution levels, particularly in developing strategies to achieve compliance with the U.S. EPA's Regional Haze Rules, which are aimed at returning visibility in protected regions to natural levels.

• Some studies have begun to estimate the premature mortalities attributable to long-range transport of PM, but they are limited by large uncertainties in estimates of imported and exported PM and by lack of a mechanistic understanding of how the individual components of PM are linked to health. Domestic sources of PM are thought to be by far the larger risk to human health, but (because there is thought to be no threshold for PM health impacts) the import of PM from distant sources could add to the health burden. Other impacts, such as perturbations to the regional radiative balance and harmful ecological and agricultural impacts of long-range PM transport, could be significant but have not been rigorously evaluated.

Mercury

• Hg is truly a global pollutant, as it has the potential, once emitted from any source, to be transformed to different chemical forms, transported through the atmosphere, and deposited long distances from the point of origin. Reservoirs that accumulate Hg include the atmosphere, the oceans, freshwater systems, soils, biota, and the cryosphere. Hence intercontinental transport is an important process that clearly affects U.S. exposures. Continued emissions will increase the amount of Hg in the global pool available for long-range transport and recycling among these reservoirs in the environment.

• As a result of increasing emissions, Hg deposition to the contiguous United States has increased since the beginning of the industrial revolution. Recent modeling studies suggest that a range of 10 to 80 percent of Hg deposited to the United States is from domestic anthropogenic sources, with an average of about 30 percent for the country as a whole. The rest is derived from natural sources and the global pool. It is clear that a component of Hg in fish (the primary route of human exposure to Hg) is derived through atmospheric deposition coming from the global atmospheric Hg pool.

• Our ability to quantify the magnitude and rates of exchange among these Hg reservoirs is limited by incomplete knowledge about atmospheric chemical processing, dry deposition, and the resuspension of deposited Hg back to the atmosphere, and by insufficient observational data needed to evaluate models.

Persistent Organic Pollutants

• There is substantial observational evidence that POPs can be transported over long distances. Transpacific atmospheric transport of POPs to the continental United States is relatively well characterized, but other potentially important transport pathways are not.

• There is evidence that atmospheric concentrations of historically used pesticide-related POPs are declining due to global regulations, while concentrations of combustion-related POPs, as well as some chemicals currently in use that have the potential to be considered POPs, are increasing due to growing emissions.

• U.S. efforts to reduce exposures to POPs are clearly impacted by long-range transport, and there is potential for the U.S. population (as well as some remote high-elevation and high-latitude U.S. ecosystems) to be exposed to increasing concentrations of certain POPs. This includes the re-release of legacy POPs by melting glaciers, forest fires, and vaporization from soils and oceans.

• At present it is difficult to quantify intercontinental transport fluxes or characterize the significance of this influence because of limited observations and modeling tools, incomplete scientific understanding of the photochemical processes that affect POPs during transport, and lack of standard national goals for POPs deposition.

All four pollutant classes had similar findings regarding expected future changes in long-range transport trends and impacts. Future climate change is likely to affect the patterns of emission, transport, transformation, and deposition for all pollution species. Predicting the net impacts resulting from all the various potential changes, however (e.g., changes in meteorology, atmospheric chemistry and dynamics, source and sink strengths), is extremely difficult with present knowledge.

Likewise, for all of the pollutants in question, future anthropogenic emissions are expected to increase in East Asia and much of the rest of the developing world, based on population and economic growth projections. In the long run these increases could be mitigated by increasingly stringent pollution control efforts and future international cooperation in developing and deploying pollution control technology and information exchange programs.

CROSSCUTTING RECOMMENDATIONS

The Committee identified a number of crosscutting opportunities to create multipollutant tools and methods that will allow better characterization (and eventually control) of long-range pollutant transport problems.

Pollutant Fingerprinting To better identify source-specific characteristics for individual pollutants and complex particles, the Committee recommends continued investment in advancing cutting-edge fingerprinting techniques (e.g., source-specific ratios of pollutant gases and particles, isotopic analyses, single-particle analysis systems) and the widespread deployment of such techniques in both ground- and air-based studies aimed at refining our assessment of pollution sources, transport, and chemical transformation. Particular emphasis should be placed on using these techniques to advance understanding of the complex heterogeneous reactions of organic species and the complexities of the mercury biogeochemical cycle.

Emission Inventories and Projections To enhance our ability to understand, forecast, and manage changing emission sources, and thus adequately model long-range pollutant transport events, the Committee recommends designing field experiments that not only confirm emission totals but also link them to the fundamental sources and processes that generate them; improving the accuracy, timeliness, spatial and temporal resolution, multipollutant coverage, and inter-comparability of Northern Hemisphere regional and national emission inventories; enhancing understanding of legacy emissions of Hg and POPs; and stimulating the collection, evaluation, and use of airborne and satellite multipollutant concentration data using inverse chemical transport modeling techniques to independently evaluate available emission inventories.

Meteorological Processes To improve our knowledge of how pollutants move between the atmospheric boundary layer (where most pollutants are emitted and eventually redeposited) and the free troposphere (where long-range transport is most efficient), the Committee recommends developing a better understanding of the basic dynamic processes involved in the entrainment, long-range transport, and deposition of pollutants through focused field studies, advances in satellite technology, improved data assimilation methods, and enhanced meteorological and transport modeling capabilities. Meteorologists and atmospheric chemists, including both modelers and measurement specialists, should collaborate on efforts to develop better measurement techniques and improved numerical models that will enable us to adequately quantify the role of distant sources on local air quality.

Ship and Air Transport Emissions It is necessary to better understand how emissions from ocean shipping and transport aircraft affect atmospheric composition, and can complicate the detection and characterization of long-range atmospheric pollutant transport from traditional land-based sources. The Committee thus recommends coordinating studies of long-range atmospheric pollution transport with studies of ship and aircraft emissions, with the goal of determining methods to distinguish among these pollutant sources in source attribution studies.

Integrated Source Attribution System Determining the impacts of distant emissions on U.S. receptor sites is a daunting task because local and regional sources usually dominate long-range transport from foreign sources. Approaching this problem one pollutant at a time is sub-optimal since most sources emit a suite of pollutants that are transported together. Further, no one type of observation or atmospheric model is likely to solve complicated source attribution problems. The Committee strongly recommends that an integrated source attribution program be established to assess the contribution of distant sources to U.S. air quality and to evaluate the effectiveness of national control strategies to meet environmental targets. The program should focus on improving capabilities within the following key areas: emissions measurements and estimates, atmospheric chemical and meteorological modeling, long-term ground-based observations, satellite remote sensing, and process-focused field experiments. These different components need to be integrated as effectively as possible to focus on source attribution applications. An expert group should be established to help design this source-attribution network (e.g., suggesting parameters to be measured, identifying appropriate monitoring sites, developing an embedded research program). These efforts should consider the need for international cooperation with opportunities to collaborate existing international efforts, such as the World Meteorological Organization's International Global Atmospheric Chemistry Observation program.

CONCLUDING THOUGHTS

The pollutants discussed in this study do not represent all species of concern, but they do illustrate the variability of pollutant composition and behavior and provide focused examples for analyzing the phenomenon of long-range pollutant transport. Present global socioeconomic scenarios predict that adverse air quality impacts from distant sources of pollution are likely to grow and cause increasing concern in the United States and other nations that are determined to provide clean air for their citizens and their ecosystems. Enhancing atmospheric observations, chemical transport models, trend analyses, the understanding of pollutant chemical and physical transformations, and emission inventories and projections will all be critically important to better quantify such effects.

The Committee wishes to emphasize that our atmosphere connects all regions of the globe, and pollution emissions within any country can affect populations, ecosystems, and climate properties well beyond national borders. Likewise, measures taken to decrease emissions in any region can have benefits that are distributed across the Northern Hemisphere. The United States, as both a source and receptor of long-range pollution, has an interest in remaining actively engaged in this issue, including support of more extensive international cooperation in research, assessment, and ultimately, emissions control efforts.

It is clear that local pollution can be affected by global sources, although in most cases air quality violations are driven by local emissions. Regardless of where the pollution originates, protecting human and ecological systems from dangerous levels of pollution should be the policymakers' primary objective. Meeting this objective will require strengthening domestic pollution control efforts to whatever levels are required to ensure that a population's total pollution exposure (from local, regional, and distant sources) does not exceed safe levels. Reducing the impacts of distant emissions on local air quality cannot be achieved by domestic efforts alone. Cooperative international action needs to be pursued vigorously to advance our understanding of long-range transport of pollution and its impacts, and to use that understanding to effectively control emissions from both domestic and foreign sources.

1

Introduction

INTERNATIONAL IMPACTS ON LOCAL AND REGIONAL AIR POLLUTION

Human activities produce or enhance the release of a wide range of airborne substances that affect air quality; and poor air quality has long been recognized as an undesirable side effect of urban population concentrations and intensive industrial and agricultural activities. By the 1950s it was widely recognized that emissions of combustion-generated exhaust gases and smoke particles, malodorous (and sometimes toxic) emissions from factory and farm operations, and windblown dust from unpaved roads, construction sites, and degraded agricultural fields could all directly degrade local air quality. By the 1960s secondary gaseous and particulate smog pollutants, formed by atmospheric reactions triggered by photochemical processing of directly emitted pollutants, were confirmed as important components of local air pollution. During the last quarter of the 20th century it became increasingly clear that emissions and associated secondary pollutants could have serious effects well beyond the local community, impacting air quality, public health, and ecosystem viability on scales of hundreds to thousands of kilometers downwind. Recently atmospheric studies have demonstrated that near-surface pollutants can be lofted to high altitudes where strong winds can transport high concentrations across oceans to other continents.

We now know that atmospheric pollutants often have long-distance impacts on regional and continental scales (acid and nutrient deposition, atmospheric haze, particulate matter climate impact, persistent toxic

11

organic pollutant contamination) as well as hemispheric and global scales (greenhouse gas warming, mercury contamination, and stratospheric ozone depletion) (NRC, 1998). As we begin the 21st century, strong links between climate change and global air quality are becoming apparent, and the possibility that intercontinental transport of air pollutants can have a significant impact on local and regional air quality is more widely recognized (NRC, 2001; HTAP-TF, 2007; Stohl, 2004).

Air pollution, once thought of as purely a local issue, now is recognized as a complex problem that is also subject to regional, hemispheric, and even global influences. Although domestic sources are the primary contributors to most of our nation's air quality problems, the United States is both a source and a receptor for pollutants transported great distances. Pollutants not only flow across our borders with Canada and Mexico but also travel between North America and Asia, Africa, and Europe. These pollutants contribute to public health threats, degraded visibility, agricultural and native vegetation injury, decreased domestic and wild animal viability, infrastructure materials damage, poorer water quality, degraded aquatic ecosystems, and climate change.

Long-range transport of air pollution from international sources is receiving increased attention in both the scientific literature and popular media. It is also an increasing concern for managers charged with meeting air quality standards. Air quality management stakeholders have advanced differing positions on how localities should address the impact of international pollutant transport in air quality planning. There also is concern about international pollution transport as a source of continuing exposure to chemicals that have been banned in the United States and other countries (e.g., certain persistent organic pollutants, POPs). The U.S. government is under increasing pressure from European countries to mitigate the transport of pollution across the Atlantic, and to address concerns about the contamination of pristine Arctic environments. Although there is now qualitative evidence of long-range transport of pollution, we need a more quantitative understanding of the true magnitude and dynamics of these flows, in order to assess their impacts on human health, ecosystem viability, and other concerns, and to design effective cooperative international control strategies. Some basic concepts needed to discuss the long-range transport of air pollutants are defined in Box 1.1.

To help address these issues the National Research Council (NRC) of the U.S. National Academies was asked to convene an ad hoc study committee to assess and summarize current understanding of the long-range transport of four key classes of pollutants: ozone (O_3) and its precursors, particulate matter (PM) and its precursors, persistent organic pollutants

BOX 1.1

Definitions of Key Terms

Because this report focuses on the impact of polluting substances from distant sources on local pollution concentrations, it is important to define some key environmental concepts and terms. An **air pollutant** is any airborne substance that has the capacity to adversely impact human health or other components of Earth's biosphere. Most air pollutants have some natural background level (discussed below) although a few, such as some POPs, are entirely anthropogenic in origin.

Air pollutants can be gaseous substances dispersed as individual molecules or very small condensed-phase liquid droplets or solid particles. The condensed airborne material is often collectively termed **aerosol particles** or **particulate matter** (**PM**). Air pollutants also are characterized by their creation mechanism. **Primary pollutants** are directly emitted into the atmosphere from pollution sources, while **secondary pollutants** are formed in the atmosphere from pollutant precursor species or other pollutants by chemical reaction. Secondary pollutants can be either gases or particles.

Precursors for specific secondary pollutants are airborne substances with the capacity to be transformed into the pollutant of interest; they may or may not be pollutants in their own right. For example, nearly all volatile organic compounds (VOCs) are precursors for the formation of ozone, an important secondary pollutant, but some VOCs like benzene or formaldehyde are also highly toxic substances, while other VOCs like ethane or propane are not dangerous at normal atmospheric concentrations. Pollutants and pollutant precursors may have sources that are classified as either anthropogenic (caused by human activity) or natural (produced in the absence of human activity). Windborne dust from desert regions is an example of a natural primary pollutant; although humans can stimulate excess dust by driving vehicles over desert soil or engaging in unsustainable agricultural practices that enhance desert formation. Formaldehyde is an example of a gaseous pollutant with multiple sources. It can be a primary pollutant emitted in motor vehicle exhaust or a secondary pollutant formed from the atmospheric oxidation of VOCs. This report focuses on ozone (O_3), a gaseous secondary pollutant, and its precursors; particulate matter (PM), which has important primary and secondary sources; mercury (Hg), which is found in both gaseous and condensed phase (PM) forms; and a number of persistent organic pollutants (POPs), most of which, like mercury, are partly gaseous and partly PM species.

continued

BOX 1.1 Continued

Long-range transport is a term that must be defined in context. In the United States, long-range pollutant transport is usually discussed in terms of multistate pollution episodes where emissions in upwind states lead to high pollutant levels in downwind states. In the context of this report long-range transport around the Northern Hemisphere generally refers to transpacific or transatlantic flows to or from North America or flows between Europe and Asia. North/south long-range transport in this report refers to pollution movement between semitropical regions and midlatitudes or midlatitudes and northern/arctic regions.

A key issue for this report is to discern when significant concentrations of pollutants from distant international sources contribute to local pollution levels. Such events can appear as pollution-rich plumes that are sufficiently discrete to be recognized by available measurement systems or by a general diffuse enhancement of relevant ambient pollutants (i.e., persistent transport). In either case the contribution from long-range transport must be defined relative to **background concentrations.** From an observational perspective, background (or baseline) concentrations are often estimated as the average weakly varying concentrations against which **pollution plumes** (events of enhanced pollutant concentrations) can be referenced. This provides a simple way to quantify the pollutant enhancements associated with plumes, although the background can still contain a contribution from a particular source diluted over the larger scale. An alternative observational definition of background is the lower envelope of the frequency distribution of concentrations, reflecting conditions of minimum pollutant source influence. This approach removes the influence of diluted pollution sources, although the definition of the lower envelope of concentrations may not be robust.

Another method used to estimate background is to conduct a sensitivity simulation in a global model, in which the pollution sources that are not considered to be part of the background are inactivated for longer than the lifetime of the pollutants being studied. This removes ambiguity in defining the background but cannot be directly verified by observations, so the validity of the model must be carefully evaluated. Finally, the U.S. Environmental Protection Agency (EPA) has defined a policy relevant background (PRB) as the level of a specific pollutant that would exist in the absence of North American emissions of that pollutant and its precursors. The PRB cannot be measured directly but can be estimated by modeling studies. Discussions in Chapters 2-5 will indicate which type of background is meant in the context of the pollutant species being discussed.

(POPs), and mercury (Hg).[1] Considerably more research effort has been expended to investigate photochemically-derived air pollutants like O_3 and PM than airborne toxic chemicals like Hg and POPs; as a result, the emission sources, atmospheric transformation, and removal processes are better characterized for the former. This is also because atmospheric concentrations of O_3 and PM are much higher than those of airborne Hg and individual POPs, thus making measurements easier to perform. Accurately quantifying the local impacts of distant emissions is difficult for all pollutants, but the less complete knowledge base available for Hg and POPs results in a greater level of uncertainty in the evaluation of long-range transport impacts on health and the environment.

This study is aimed at aiding U.S. decision makers engaged in developing domestic and international air quality management policies, as well as those engaged in planning and funding atmospheric and environmental research. We hope our findings and recommendations will help to shape both near-term policy responses and the research investments that are needed as a foundation for designing effective longer-term policy responses. It also is intended to help inform the final assessment by the Long Range Transport of Air Pollutants (LRTAP) Task Force on Hemispheric Transport of Air Pollution (HTAP-TF). The HTAP efforts are not focused specifically on pollution flow into out of the United States but rather on examining pollutant flows throughout the Northern Hemisphere. We anticipate that our assessment effort, focused specifically on the information needs of U.S. policy makers, will be a valuable complement to the HTAP assessment efforts. The NRC study was designed to be independent of the HTAP process, but several steps were taken to ensure effective coordination and communication between the two efforts. For example, several of the NRC committee members are directly involved in HTAP analysis and assessment activities, the first NRC committee meeting was held in conjunction with a workshop of the HTAP task force, and several of the speakers invited to subsequent NRC meetings were active in the HTAP process.

MOTIVATIONS FOR CONCERN

Despite increasing growth and urbanization, air quality in developed countries generally has improved over the past 25 years because of serious efforts to set and meet air quality standards. However, air quality in much of the less developed world has declined due to increasing emissions from rapidly expanding and poorly regulated motor vehicle fleets, growing industrial and power generation activities, and domestic coal and biomass

---

[1] Information on the committee's sponsors, its schedule, and its full statement of task are presented in Appendix A.

burning. Widespread overuse and degradation of crop, grazing, and wood-lot lands also have led to greater levels of airborne dust; more intensive agricultural activities have led to increasing levels of fixed nitrogen and persistent organic pesticides.

Maintaining or improving living standards for a rapidly growing and urbanizing world population tends to increase global pollutant emissions from the industrial, transportation, agricultural, and energy sectors, as well as domestic and commercial heating, cooling, lighting, and cooking systems. Increased emissions push ambient concentrations of both primary and secondary air pollutants higher. Scientific studies have quantified the adverse impacts of atmospheric pollutants on human health, agricultural yields, and natural ecosystems. These studies often reveal serious impacts at relatively low pollutant concentrations, putting pressure on air quality managers to decrease the allowable concentrations of many air pollutants.

The conflict between increasing air pollutant levels in some parts of the world, and more stringent air quality standards in other parts of the world, leads to concerns that if even small fractions of the pollutants from Nation 1 reach Nation 2, Nation 2 may find it significantly more difficult to meet its mandated air quality goals. The cost of not meeting air quality standards on human health, agricultural production, and ecosystem viabilty can be high, so accepting higher pollution levels is an unattractive option. A brief overview of some of the costs associated with poor air quality follows.

Human Health Impacts   Alleviating the impact of air pollution on human health is the primary motivation of air quality management programs around the world. Some air pollutants can affect human health directly by inhalation, introducing pollutants that directly harm the throat and lungs and reach other internal organs after entering the pulmonary blood flow. Air pollutants also can be deposited to the ecosystems that provide humans with food and drinking water, including agricultural soils, fresh surface and ground water, and marine waters, where they enter the food chain or drinking water supplies, leading to human exposure by ingestion. Of the four classes of pollutants considered in this report, human exposure to O_3 and PM is dominated by inhalation, while exposures to POPs and Hg occur primarily by ingestion.

High concentrations of ozone and particulate matter increase the risk of cardiovascular and respiratory diseases, as well as a wide range of other adverse health outcomes (Cohen et al., 2004; EPA, 2004, 2006a; WHO, 2006; NRC, 2008). The total health burden caused by these pollutants can be considerable; the World Health Organization estimates that 41,200 Americans die prematurely each year due to exposure to elevated concentrations of PM that is less than 10 microns in diameter (WHO, 2002). Efforts to reduce airborne particulate matter can produce dramatic results;

a recent correlation of mortality data in 51 metropolitan areas with reductions in fine PM (less than 2.5 micrometers in diameter) between the late 1970s and early 1980s and the late 1990s and early 2000s estimated that reducing the average fine PM ambient concentrations by $10\mu g/m^3$ resulted in increasing mean average lifetimes by 0.61 ± 0.20 years (Pope III et al., 2009). A number of time-series and cohort studies have found the concentration–response relationship between PM exposure and mortality response to be close to linear with no evidence of a threshold below which no adverse health impacts are observed (Daniels et al., 2000; Pope, 2000; Schwartz et al., 2002, 2008; Pope and Dockery, 2006).

The impact of airborne ozone on human mortality is not as large as that of fine PM, but a recent NRC review confirmed that short-term exposure to high-ambient ozone levels does produce significant premature mortality, and that the risk of mortality is not limited to those already at a high risk of death (NRC, 2008). The total health burden of PM and ozone pollution is considerably larger than their impacts on premature mortality alone and is dominated by subclinical and symptomatic events. The risk of suffering an adverse health event due to exposure to air pollution is small, but widespread exposures result in large numbers of affected individuals. The extent of these health impacts depends on dose, with susceptibility varying with age, health status, diet, and genetics.

Mercury is a neurotoxin that can cause hand tremors, increases in memory disturbance, and other adverse health impacts (EPA, 1997). There is evidence of adverse human health effects from environmental exposures to POPs, and significant concern about elevated concentrations of these chemicals in a range of tissues, including venous and cord blood, adipose tissue, and breast milk (Li et al., 2006a).

Environmental Impacts Although impacts on human health may be the dominant reason for concern about air pollution (coming from both local and nonlocal sources), there are a variety of other adverse impacts that underlie interest in air pollution dynamics. These include impacts on ecosystem health, including impacts on agriculture productivity and wild fish populations, which in turn have serious implications for human well-being; reduced visibility due to pollution-induced haze, which obscures views in scenic locations around the world; and degradation of materials in buildings, monuments, vehicles, and other infrastructure components. The effects of short-lived air pollutant species on regional and global climate, through both direct interaction with atmospheric radiation and indirect effects related to changes in cloud properties are a growing concern. More information about specific air pollutants and other environmental impacts is presented in the following four chapters.

LONG-RANGE TRANSPORT OF POLLUTION

The fact that pollutants emitted near the surface of one continent can influence human health or crop yields on another continent many thousands of miles away seems unlikely to many people. One might suppose that as polluted air mixes with relatively clean air over the oceans, the pollution concentration would be diluted to inconsequential levels. In some instances, however, surprisingly intense pollution plumes can travel across broad oceans, delivering significant pollutant concentrations to a downwind continent. Under other conditions (for some pollutants) diffuse export can lead to an overall increase in tropospheric abundance, thus increasing pollution in background clean air. In all cases the ultimate impacts will depend on how pollutants being transported aloft are mixed down to the surface. A brief overview of the dynamics of atmospheric air masses and their associated time and distance scales is presented below, and a more detailed explanation is provided in Appendix B.

Atmospheric Dynamics of Long-Range Transport   Most meteorological phenomena that affect long-range pollutant transport occur in the troposphere, the lowest portion of the atmosphere, which extends from the surface to ~ 18 km in the tropics and ~ 8 km in polar regions. The tropopause is the boundary between the troposphere and the stratosphere, the next higher atmospheric layer. The troposphere can be subdivided into two regions—the planetary boundary layer (PBL) and the free troposphere. The PBL typically is confined to the lowest 1-2 km of the troposphere. Although most pollutants are released in this layer, their horizontal transport usually is quite slow due to relatively weak winds near Earth's surface. Conversely, pollutants that are transported from the PBL into the free troposphere often travel great distances because of the stronger winds aloft, including the jet streams, the strongest rivers of atmospheric flow that encircle Earth. The mechanisms producing upward transport into the free atmosphere (e.g., thunderstorms and middle latitude low-pressure systems) play important roles in determining whether or not long-range pollutant transport will occur. The weather phenomena that affect the long-range transport of pollution range in size from small, short-lived turbulent eddies to systems that span continents and can last weeks or months. Thunderstorms, sea breezes, and high and low-pressure systems all play a role in transporting pollutants both horizontally and vertically.

It is convenient to think of intercontinental transport in terms of two different mechanisms. In the first, transport occurs in well-defined puffs of pollution that are lifted out of the polluted PBL of the source region. Once in the free troposphere relatively strong and persistent winds move the polluted air mass great distances with minimal dilution and removal of its constituents (except during rain events) although the pollutants eventually become diluted into the background. This type of transport has been documented frequently. The air mass then can be mixed down to the surface, where it contributes to local air quality degradation. Because the polluted air mass remains relatively distinct, it is possible to identify its impacts from observations, even at great distances. In the second transport mechanism, pollutants from the source region are rapidly mixed into the background air. When this occurs, the source of the pollutants is more difficult to trace, and the exact contribution from each region is more difficult to determine. These two transport mechanisms occur simultaneously to varying degrees. Chemical transport models (CTMs) are important tools to quantify the contributions for both types of transport.

The overall transport of pollutants in the free atmosphere of the middle latitudes is from west to east due to the prevailing westerly winds (Figure 1.1). This easterly transport is greatly influenced by transient low-pressure systems. The east coast of Asia is an area of enhanced low-pressure development where systems form every few days and travel eastward toward North America. These transient lows contain several major transport channels, with the warm conveyor belt being the best defined. This conveyor belt begins near the surface in advance of cold fronts. It can transport surface-based pollution from the major industrialized areas of eastern Asia northward and then eastward. Air slowly rises out of the boundary layer, often

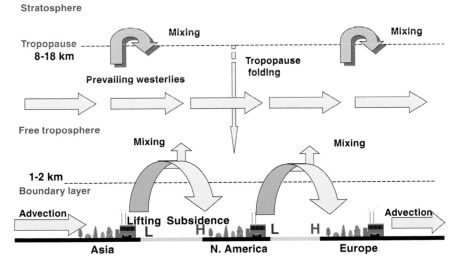

FIGURE 1.1 Schematic of the dominant dynamical processes involved in long-range midlatitude pollution transport. Ground level H and L symbols represent high- and low-pressure systems.

reaching the upper troposphere where the pollution is rapidly transported eastward toward North America by the strong winds aloft. West and south-west of the low in the advancing cooler air mass, dry air from the upper troposphere and lower stratosphere subsides toward the lower levels. This area often contains relatively small regions of a lowered tropopause (folds) and other mechanisms that mix stratospheric air into the troposphere.

The second major area of low-pressure development is the east coast of North America. The structure of these systems is similar to those form-ing near eastern Asia. However, these eastward moving storms transport surface-based pollution from the industrialized North American east coast to higher altitudes and then toward Europe. Europe generally is not an area of frequent low-pressure storm development. Winds within both the bound-ary layer and free troposphere climatologically are from the west, thereby completing the global transport cycle. At lower latitudes the prevailing winds typically flow from east to west, while the prevailing midlatitude flows are from west to east; winds from northern Africa can bring Saharan dust to the Caribbean, Florida, and other Gulf Coast states. Figure 1.2 illustrates a number of relevant long-range transport paths and representa-tive time scales. Note that timescales may differ for concentrated pollution plumes and general transport of background air; for instance, Liu and

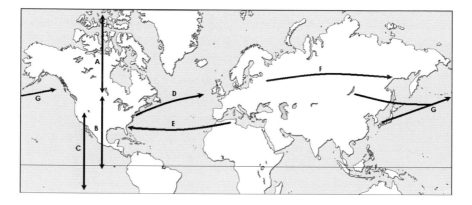

FIGURE 1.2 Major atmospheric transport pathways affecting North America. The general timescales of transport estimated by the committee from trajectory studies and other sources are: (A) Midlatitudes–Arctic exchange: 1-4 weeks. (B) Midlatitudes–Tropics exchange: 1-2 months. (C) Northern Hemisphere–Southern Hemisphere exchange: ~ 1 year. (D) North America to Western Europe: 3-13 days. (E) Northern Africa to North America: 1-2 weeks. (F) Eastern Europe to Asia: 1-2 weeks. (G) Eastern Asia to North America: 4-17 days. For (A), (B), and (C), transport occurs in both directions, depending on altitude (see Appendix B for details).

Mauzerall (2005) estimate that intercontinental transport times are, on average, 1-2 weeks longer than rapid transport times observed in plumes. Appendix B provides further discussion and illustration of these concepts.

Long-Range Pollutant Transport Example Figure 1.3 shows an example of transport from Asia to the Mt. Bachelor Observatory in central Oregon. On April 25, 2004, observations at Mt. Bachelor detected enhanced concentrations of carbon monoxide (a long-lived pollution plume tracer), ozone, fine submicron particulate matter, elemental mercury, as well as particulate-phase polycyclic aromatic hydrocarbons (PAHs) and hexachlorocyclohexane, which are persistent organic pollutants. The enhancements in the observed pollutants occured nearly simultaneously. The source for these pollutants can unambiguously be attributed to Asia based on meteorological data, the chemical pattern of pollutants and the absence of any short-lived tracer pollutants (e.g. NO_x with lifetime τ of less than 1 day) that would indicate North American sources (see Jaffe et al., 2005a; Weiss-Penzias et al., 2006, 2007; Primbs et al., 2008a,b).

Meteorological analyses indicate that the polluted air mass detected at Mt. Bachelor on April 25 and 26 took approximately eight days to travel from East Asia to central Oregon.

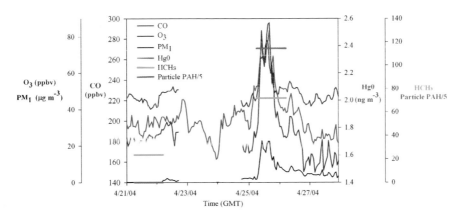

FIGURE 1.3 April 2004 observations from the Mt. Bachelor Observatory in central Oregon (2.7 km above sea level). On April 25, an air mass containing a variety of pollutants emitted in Asia was identified at the station. Note that measurements of HCH and PAHs occur on discrete filter samples, whereas the other pollutants are measured continuously. Units are pg/m^3 for the 24-hr integrated sample measurements for April 21-22 and 25-26 of particulate phase PAHs and HCH (values shown on right axis).

Chemical and Microphysical Transformations During Long-Range Transport Pollutant species have finite residence times in the atmosphere before they are chemically transformed or deposited to the surface. In both the atmospheric boundary layer and the free troposphere, photochemical reactions can transform many pollutants into other chemical species. Microphysical processes such as particle nucleation, vapor condensation, and semivolatile evaporation transfer species between the gas phase and particulate condensed phases. Wet deposition can remove pollutants absorbed into cloud droplets or entrained into falling precipitation. Dry deposition (of both gaseous and particulate pollution) to vegetation, soil, water and humanmade material surfaces can occur for pollutants in the near-surface boundary layer. All of these transformation and removal processes (some of which are depicted in Figure 1.4) can operate during long-range trans-

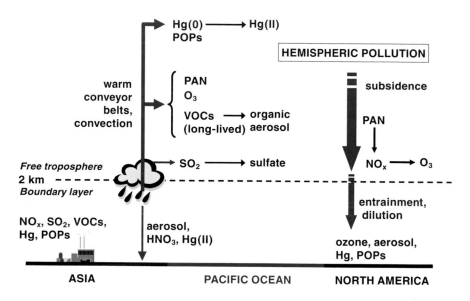

FIGURE 1.4 Some basic chemical and microphysical mechanisms involved in long-range (transpacific) pollution transport. Chemical transformation of Hg(0) to Hg(II) allows reactive Hg to be absorbed by aerosol particles and cloud droplets, peroxy-acetylnitrate (PAN) can sequester reactive nitrogen oxides (NO_x) in the cold free troposphere but releases it to form ozone when air masses subside and warm. SO_2 is oxidized to sulfate, and volatile organic compounds (VOCs) are oxidized to less volatile organic species by both gas phase and cloud droplet reactions, forming additional particulate matter. Wet deposition scavenges aerosol PM, nitric acid formed by oxidizing NO_x, and Hg(II), while dry deposition removes a wide range of gaseous species as well as particle-bound species. The diagram depicts processes occurring during transpacific transport; these same processes occur during transatlantic transport.

port of air masses, with chemical transformations significantly reducing many reactive species with atmospheric chemical lifetimes of less than a week (Finlayson-Pitts and Pitts, 2006). In general, long-range transport can be significant when pollutant transformation rates are slow relative to long-distance transport times. More detailed examples of chemical transformation and removal processes are presented in subsequent chapters that discuss the production and transport of specific pollutant classes.

Tools for Understanding Long-Range Transport It is challenging to detect and quantify the impact of long-range transport on local air quality that is dominated by local sources. Clear examples of the impact of well-organized long-range transport of plumes like that described above are most common for free tropospheric observations; boundary layer examples are less common. In general, combinations of sophisticated measurement methods and advanced CTMs are required to recognize and quantify the boundary layer and surface impacts of long-range transport events. Models are also used to predict how future emission levels in various geographic source regions will impact pollutant concentrations in distant receptor regions.

Some of the currently utilized measurement and modeling tools are described in Boxes 1.2 and 1.3, respectively. Specific results obtained by utilizing these measurement and modeling tools are presented in subsequent chapters describing long-range transport of relevant pollutant classes. A prospective advanced integrated measurement and modeling system is presented in Chapter 6.

Emission Inventories Requirements Long-range transport of pollution is directly dependent on the magnitudes of emission fluxes of primary pollutants and precursors in upwind source regions. Accurate, up-to-date knowledge of the magnitudes of primary emissions, as well as their temporal and spatial distributions, are required to evaluate the range of impacts on air quality, climate, ecosystem viability, and human health for downwind receptor regions. Some basic concepts necessary to classify and quantify pollutant and pollutant precursor emissions are presented in Box 1.4.

Adequate emission inventories (EIs) are critical inputs for the chemical transport models that are traditionally used to quantify receptor region concentrations of both primary and secondary pollutants. These pollutants have a wide range of atmospheric lifetimes; high spatial and temporal fidelity predictions of impacts for species with lifetimes shorter than or approximately equivalent to transport times require similarly resolved emission data.

A recent three-nation assessment of North American emission inventories identified a wide range of deficiencies in EIs that are inputs to CTMs designed to predict photochemical oxidant and fine particulate concentra-

---

BOX 1.2

**Observational Tools Used In the Study of
Long-Range Transport of Pollution**

The study of long-range transport requires an integrated observing strategy encompassing techniques that span a large range of spatial and temporal scales. Satellite observations provide large-scale multiyear records of a limited number of key atmospheric trace gas and aerosol parameters. They produce important but limited resolution data in the horizontal, vertical, and temporal domains. However, these measurements provide critical context for more localized observations, and help extend local measurements to regional and global scales.

Local and regional-scale atmospheric chemistry and dynamics are readily observable from aircraft with integrated chemical and meteorological instrumentation. Intensive airborne campaigns can provide a wide range of measurements, a detailed view of the important atmospheric processes, and good spatial and temporal resolution pollutant concentration measurements. Ship, train, and van-borne mobile laboratory instrument suites can produce similar data for the lower boundary layer, but these studies generally have limited temporal and spatial coverage. Surface monitoring sites and vertical profiling instruments yield datasets that are specific to a single location. These datasets are particularly useful for discerning trends as they are the only long-term, semicontinuous measurement of atmospheric pollution that we have.

---

tion fields (NARSTO, 2005). These deficiencies are present even in the national and regional EIs produced by relatively experienced air quality management agencies in the three North American nations. This same study noted that assessing the adverse impact of long-range transport of pollution on North American air quality would require much improved emission inventories for the entire Northern Hemisphere.

Emissions from regions with less experience in developing EIs are widely recognized to have even more serious problems, caused both by lack of basic emission data and inadequate institutional support for gathering and processing required data. EIs for regions with rapidly changing economies are particularly problematic. In addition to uncertainties in current emissions, high rates of economic growth in these regions are accompanied by significant population shifts and rapid changes in technology and infrastructure, both of which produce rapid changes in emission quantities and distributions. Thus, even if EIs are relatively accurate when compiled, they

BOX 1.3

**Modeling Tools Used In the Study of
Long-Range Transport of Pollution**

Chemical transport models (CTMs) are important tools that are used to explore pollution transport pathways and to assess the impact of long-range transport on ambient pollution levels. Identification of effective measures to reduce pollution loads at a specific place requires quantification of contributions from local and distant sources, distinguishing between natural vs. humanmade sources as well as among local, regional, transboundary, and intercontinental sources. We are seldom confronted with a situation where the observed concentration or deposition fields at a particular location are due solely to emissions from a particular source. The further we are from source regions and the longer the atmospheric lifetime of the pollutant, the more likely it is that multiple sources contribute to observed concentrations. Since observations generally cannot answer such questions on their own, models are widely used to provide the required information.

CTMs are computer models that link the emissions of pollutants to their ambient distributions. They integrate the meteorological, chemical, and physical processes that control the fate and transport of pollutants in the atmosphere. Over the last decade our ability to predict air quality has improved significantly due to advancements in our ability to measure and model atmospheric chemical, transport, and removal processes. CTMs now track scores of chemical species in various physical phases interacting with hundreds of chemical reactions. In addition, the transport components of CTMs now are run in close interaction with dynamic meteorological models. Computational power and efficiencies have advanced to the point where CTMs can simulate pollution distributions in an urban airshed with a spatial resolution of less than one kilometer, and can cover the entire globe with horizontal resolution of less than 100 kilometers.

Assessing the impacts of distant sources on local air quality requires methods to track the contributions of specific source regions to pollution levels over regions of interest (i.e., the source-receptor relationships). A source-receptor (S-R) relationship quantifies the contribution of an emission source to the concentration or deposition of a specific pollutant at a receptor point elsewhere. While CTMs have been used to calculate S-R relationships over relatively short distances (hundreds of kilometers), their application in global S-R analysis is recent (HTAP-TF, 2007).

Calculations of S-R relationships can be classified into source- and receptor-oriented approaches. Most commonly used is a source-oriented

continued

BOX 1.3 Continued

approach, wherein emissions from individual source regions are perturbed (or tagged), and these perturbations are propagated forward throughout the modeling domain to future times (Wang et al., 1998; Wild et al., 2001; Derwent et al., 2004; Auvray and Bey, 2005; Sudo and Akimoto, 2007; Liu et al., 2007, 2008, 2009; Fiore et al., 2009; Saikawa et al., 2009). Calculated concentrations from a tagged source region provide that source region's footprint, which predicts the concentration from that source at all possible receptor regions. In this approach the perturbations from each source region of interest are calculated individually, allowing estimation of the individual source contributions to the concentrations at a receptor region of interest.

In the receptor-oriented approach the perturbation in the concentrations at the receptor location due to specified emissions are traced backward in time. Receptor-oriented source-receptor calculations can be performed using adjoint sensitivities as demonstrated in Hakami et al. (2006) and Henze et al. (2008), or by running air-parcel models backward in time (see Stohl et al., 2002). A key source-receptor calculation consideration is its sensitivity to the size of the perturbation. Since CTMs generally represent nonlinear systems, the response in principle should be dependent on the size of the perturbation.

S-R relationships are model constructs, currently with no simple means of verification or validation. The accuracy and validity of the S-R relationships depend on the adequacy and completeness of the atmospheric models from which they have been derived. The capabilities of the current generation of CTMs are evaluated through direct comparisons of the predicted concentration fields with observations and through multimodel intercomparison studies (Fiore et al., 2009).

Models are thus indispensible tools for advancing our knowledge of long-range transport; however, our confidence in model predictions is constrained by numerous sources of uncertainty in the basic inputs and assumptions that drive the models. This includes uncertainties in the emissions of pollutants, the meteorological process that drives pollutant transport and deposition, and the chemical and kinetic processes that drive pollutant formation and transformation. More specific details about the nature of these different uncertainties are discussed in the following chapters. Current CTMs are generally not configured to produce quantitative uncertainty estimates based on the uncertainties in these various inputs and assumptions; unfortunately, our ability to report quantitative uncertainty estimates is quite limited.

BOX 1.4

Pollutant Emissions Overview

Emissions of the pollutants and pollutant precursors described in this report may be associated with both natural and anthropogenic (human-caused) sources.

Energy and Economy. We use the term "energy and economy" to describe sources related to energy consumption and economic activity. These sources, which are the most directly controllable, include emissions from energy production, industry and consumer products, homes, and agriculture. International ships and aircraft are a rapidly growing portion of these emissions; regulatory standards for these emissions are established by international management activities rather than by local air quality agencies. In the future, international economic activity will probably increase, but new and cleaner technology can decrease the amount of emissions per activity. Emission standards that drive adoption of clean technology may be implemented, usually after a country has enough capital to invest in its environment. In the interim, air quality in developing countries can be extremely poor (Gurjar et al., 2008). The net result of growth vs. cleanup might result in an inverted U shape for emissions: an initial increase followed by a decrease (Selden and Song, 1994), but it is not known whether improvements in technology will always counteract growth. These factors lead to uncertainty in predicting the future magnitude of air pollutant emissions. Energy and economy sources are responsible for large fractions of the ozone, mercury, and particulate matter budget, and most POP emissions.

Natural environment. Another major source of pollutants is the natural environment, a blanket term for a variety of disparate sources. Open fires are large sources of ozone precursors, PM, and some combustion-derived POPs. They may be caused by deforestation, natural causes, prescribed fires for land management, and seasonal fires for agricultural land clearing. Dust emissions contribute PM, and are affected by wind speed and the nature of the land surface. Emissions from vegetation are termed "biogenic"; complex biogenic organic gases promote ozone formation and lead to secondary organic PM, while organic biogenic particles contribute to PM loadings. Other environmental sources affect individual species, such as mercury emissions from volcanic activity and ozone incursions from the stratosphere. Many of these sources are episodic, seasonally variable, and dependent on climate. Because some of these emissions, especially biomass fires and airborne dust, are enhanced by human interactions with the natural environment, we choose not to identify these as either fully anthropogenic or fully natural.

continued

BOX 1.4 Continued

Legacy emissions. Once released to the environment mercury can cycle between elemental and molecular forms, but its natural removal requires geological time scales. Persistent organic pollutants may break down very slowly, even after they are removed from the atmosphere. For these long-lived species relofting of accumulated material can contribute a significant fraction of the atmospheric burden. Legacy emissions can result either from continuing processes such as soil resuspension or from large, episodic perturbations like forest fires. Current and future emissions of legacy material therefore depend on the history of emissions (sometimes over decades); the potential for degradation over time; the rate of exchange between terrestrial, aquatic, and biotic surfaces with the atmosphere; and their potential for long-term sequestration in soils and sediments. These emissions may be considered either anthropogenic or natural, depending on the original source.

may be badly outdated by the time they are updated, which may occur only at intervals of five to ten years.

Global emission inventories have been developed to model and assess large-scale air quality and climate processes and trends. Such inventories may also provide boundary conditions for urban or regional air quality modeling studies, but the quality of these EIs may be uneven. For example, Butler et al. (2008) assessed the emissions of CO, NO_x and nonmethane VOCs in three leading global EIs for 32 large cities and concluded that these compilations often assign inconsistent emissions levels, frequently varying by a factor of two for individual cities.

Without accurate emission inventory data for both local and upwind emissions at required spatial and temporal resolutions, CTMs cannot be expected to correctly predict how pollutants will be transformed during long-range transport or to quantify either the absolute or relative impact of these pollutants on local air quality. Deficiencies in hemispheric and global emission inventories that affect assessments of long-range transport will be discussed in subsequent chapters.

LONG-RANGE TRANSPORT POLICY CONTEXT

To assess the significance of long-range pollution transport on the achievement of environmental goals, it is necessary to identify the specific environmental goals of greatest concern and to understand the landscape of relevant national and international policies, statutes, and agreements. A general overview of this landscape is provided here. Specific policy contexts for each relevant pollutant class will be further discussed in Chapters 2-5.

International Policy Context International action on long-range pollutant transport has revolved largely around the U.N. Long Range Transport of Air Pollution (LRTAP) convention, which was signed in 1979 by 30 European countries, the United States, and Canada. The convention itself demands that the signatory parties attempt to reduce transboundary air pollution. It has been supplemented by eight successive protocols (see Box 1.5) that set reduction goals for specific compounds, and established a mechanism for supporting scientific data gathering to monitor compliance and progress. The LRTAP process has met success in achieving targeted reduction goals for specific pollutants, and has been able to adjust goals as scientific understanding evolves. LRTAP working groups continue to consider new protocols and investigate new subjects of concern.

BOX 1.5

LRTAP Protocols

- The 1999 Protocol to Abate Acidification, Eutrophication, and Ground-level Ozone. 24 parties. Entered into force on May 17, 2005.
- The 1998 Protocol on Persistent Organic Pollutants. 29 parties. Entered into force on October 23, 2003.
- The 1998 Protocol on Heavy Metals. 29 parties. Entered into force on December 29, 2003.
- The 1994 Protocol on Further Reduction of Sulphur Emissions. 28 parties. Entered into force August 5, 1998.
- The 1991 Protocol Concerning the Control of Emissions of Volatile Organic Compounds or their Transboundary Fluxes. 23 parties. Entered into force September 29, 1997.
- The 1988 Protocol Concerning the Control of Nitrogen Oxides or their Transboundary Fluxes. 32 parties. Entered into force February 14, 1991.
- The 1985 Protocol on the Reduction of Sulphur Emissions or their Transboundary Fluxes by at least 30 per cent. 23 parties. Entered into force September 2, 1987.
- The 1984 Protocol on Long-term Financing of the Cooperative Programme for Monitoring and Evaluation of the Long-range Transmission of Air Pollutants in Europe. 42 parties. Entered into force January 28, 1988.

One evolving issue is the global or hemispheric nature of air pollution transport. Because air quality gains made by the original LRTAP parties could be reversed as industrial and mobile emissions increase in nonmember countries, there is growing consensus that more effective agreements to address pollutant emissions on a global (or more specifically, on a northern hemispheric) scale are needed. LRTAP has established a Task Force on Hemispheric Transport of Air Pollutants (HTAP-TF) that is charged with advancing and assessing the state of knowledge with respect to the flows of air pollutants across the Northern Hemisphere to inform future policy negotiations under the Convention. A number of other existing bilateral, regional, and global agreements are potential vehicles for addressing the issue of long-range pollution transport, including

- the U.S.-Mexico Border 2012 Program: LaPaz agreement on cooperation for the protection and improvement of the environment in the border area (addressing O_3 and PM);
- the U.S.-Canada air quality agreement (addressing O_3 and PM) and strategy for the virtual elimination of persistent toxic substances in the Great Lakes (addressing Hg and POPs);
- the Commission on Environmental Cooperation's North American Regional action plan (addressing Hg and POPs); and
- the UNEP Stockholm Convention on POPs and Global Hg Partnership.

The UN Framework Convention on Climate Change (UNFCCC) focuses on long-lived greenhouse gases like CO_2, CH_4, and N_2O. It has long been recognized that shorter-lived radiatively active gases and particles (such as ozone and black carbon aerosols) also play important roles in climate change. As discussed in later chapters, because of the complex chemistry and physics involved, there are large uncertainties in quantifying these impacts. The Intergovernmental Panel on Climate Change and the U.S. Climate Change Science Program are making progress in including short-lived gases and particles in climate change assessment studies, but they are not presently included in the UNFCCC or other climate change mitigation agreements.

As discussed further in later chapters, studies of international pollution sources must account not only for the emissions from particular countries and regions but also from mobile sources that are not easily ascribed to any particular nation, like shipping and aviation. The International Maritime Organization recently passed a new international protocol for gradual reduction of allowable ship SO_x and NO_x emission that will enter into force in 2010. The International Civil Aviation Organization has technology development and assessment programs addressing aircraft NO_x emissions, but there are presently no international protocols or agreements for actually controlling these emissions.

Domestic Policy Context   Most national air quality goals are defined in various provisions of the Clean Air Act (CAA). Of particular interest are the National Ambient Air Quality Standards (NAAQS) (in CAA§109) for O_3, PM, and their precursors, SO_x and NO_x. The NAAQS are set by the EPA "to protect public health, allowing an adequate margin of safety, and to protect the public welfare from any known or anticipated adverse effects." The NAAQS are periodically revised in response to new scientific understanding of these impacts. For example, the 8-hr ozone standard was recently tightened from 0.08 ppm to 0.075 ppm; and the 24-hour fine particle standards were tightened from 65 $\mu g/m^3$ to 35 $\mu g/m^3$.

Meeting these more stringent air quality standards will likely require reductions in a variety of emission sources in many U.S. regions. As emissions in many other parts of the world increase, both the relative and absolute contributions of international transport to air quality problems in the United States are expected to increase.[2] Of particular interest is that incremental contributions of these flows to U.S. air quality degradation may be of the same order of magnitude as the incremental air quality improvements that are expected to result from some of the recent tightening of the NAAQS.

Improving our understanding of long-range pollution transport is particularly important for a few aspects of NAAQS implementation, for example, the concept of Policy Relevant Background (PRB), the ozone concentrations that would exist in the absence of North American anthropogenic emissions. The PRB (which cannot be directly measured but must be estimated in modeling studies) is used by EPA as a floor for estimating risk associated with alternative NAAQS levels.

Another example is CAA §179b (International Border Areas), which mandates that state, local, and regional authorities will not be penalized or otherwise burdened and held responsible for the impact of pollution emissions from foreign sources. The U.S. Chamber of Commerce and other interested parties have complained that EPA has provided no clear, consistent guidance to state, local, and regional authorities seeking to account for the impact of foreign emissions in calculating attainment of CAA standards. Also of interest is the CAA provision (319b) covering "exceptional events," which are designated as unusual or naturally occurring events (often from major forest fires) that can affect air quality but are not reasonably controllable using techniques that local air agencies may implement. The EPA

---

[2] The subsequent chapters include consideration of ambient air quality standards of other countries and WHO global-scale recommendations.

issued a rule effective as of May 2007 that governs the review and handling of air quality data influenced by exceptional events. Other CAA provisions relevant to the issue of long-range pollution transport include

- §115: International Air Pollution;
- §160: Prevention of Significant Deterioration;
- §169a: Visibility Protection for Federal Class I Areas;
- §401: Acid Deposition Control;
- §103(j): National Acid Precipitation Assessment Program;
- §112f: Hazardous Air Pollutants "Residual Risk Standard"; and
- §112m: Great Waters Program and Hazardous Air Pollutants.

Beyond the CAA the EPA recently has developed and implemented a number of more integrated multipollutant and multistate air quality control measures. This includes the NO_x SIP Call, a market-based cap-and-trade program that has successfully reduced NO_x emissions from power plants and other large combustion sources in the eastern United States. This also includes the Clean Air Interstate Rule (CAIR) issued in March 2005. CAIR was designed to further reduce emissions of sulfur dioxide (SO_2) and nitrogen oxides (NO_x) across 28 eastern states, using an emissions trading system, in order to address the effect of an upwind state's emissions on a downwind state's ability to meet air quality standards for O_3 and PM. In February 2008 a U.S. Court of Appeals issued an order to vacate CAIR, stating that it violated the EPA's statutory authority. However, in December 2008 the court withdrew this vacatur order and held that the rule could stay in effect at least until it is replaced by a rule consistent with the earlier opinion. EPA's Clean Air Mercury Rule (CAMR), the first federally man-dated requirement that coal-fired electric utilities reduce their emissions of mercury, was finalized in March 2005, but (as with CAIR) the rule was vacated by a U.S. Court of Appeals in February 2008. As of this writing, the rule has not been reinstated and its future is uncertain. By the end of 2007, 23 states had proposed or adopted Hg emission rules more stringent than those embodied in CAMR; and in mid-2008 nearly 20 states were planning to institute state-enforced Hg emissions control requirements (Milford and Pienciak, 2009).

The main exposure pathway for Hg and some POPs is through aquatic ecosystems (i.e., fish consumption rather than through direct atmospheric inhalation). Control of these species is addressed through a number of other national and state-level statutes, including the Clean Water Act; state-level Water Quality Standards and Total Maximum Daily Load (TMDL) standards; the Federal Insecticide, Fungicide, and Rodenticide Act; and the Toxic Substances Control Act. In addition, the EPA and the Food and Drug Administraiton issue consumption advisories related to Hg contamination.

The use of critical loads to assess and limit the deposition of acid and nutrient compounds also may be relevant. The concept of critical loads is currently used in Europe in the context of establishing LRTAP emission control standards for SO_x and NO_x. Although it is still a nascent concept in U.S. environmental policy, there are a number of state and provincial efforts to apply critical loads strategies for various regions of North America.

REPORT ORGANIZATION

This chapter has presented an overview of the basic phenomena controlling long-range pollution transport and the tools used to study this phenomenon. The policy context that frames our need to understand this issue also has been reviewed. Chapters 2-5 provide more detailed discussions about what is known and what needs to be learned for each of the main pollutant classes that we were asked to address in this study (ozone and its precursors, particulate matter and its precursors, gaseous and particulate mercury, and persistent organic pollutants). Chapter 6 addresses a number of crosscutting issues that affect all of these pollutants, and provides a synthesis of key findings and recommendation for future research and observational needs. Appendix A presents the committee's statement of task and some details about its sponsorship and operations. Appendix B provides a more detailed discussion of the meteorological dynamics that affect long-range transport of pollution. Appendix C presents a summary of recent observational activities relevant to long-range transport.

2

Ozone

GENERAL INTRODUCTION ON OZONE

Air Quality Ozone (O_3) is ubiquitous in the air we breathe. It is a health hazard to sensitive individuals, with a threshold (i.e. an ambient atmospheric concentration above which adverse health effects may occur), if one exists, that is below current U.S. and international standards (Bell et al., 2006). O_3 reduces lung function and thus contributes to premature deaths, emergency room visits, and exacerbation of asthma symptoms; it damages food crops and ecosystems; and it deteriorates materials. Nations have developed Air Quality Standards (AQS) for ground-level O_3 abundances to protect people and crops (e.g., the U.S. NAAQS, the Canada Wide Standard, EU standards, and WHO standards, see Figure 2.1). AQS are evolving, and in Europe have been enhanced to include protection for agricultural crops. The U.S. secondary (crops) and primary (human health) standards are the same, and but are becoming stricter as a result of increasing knowledge about the damage caused by O_3.

For each year over the period 2000–2002, between 36 and 56 percent of ozone monitors in the United States failed to meet the then current ozone standard of 0.08 ppm (Figure 2.1) for the fourth-highest annual maximum 8-hr ozone concentration (Hubbell et al., 2005). Using health impact functions derived from the published literature, they estimate that if the standard had been attained, roughly 800 premature deaths, 4500 hospital and emergency department admissions, 900,000 school absences, and > 1 million minor restricted-activity days (per year averaged over the three years studied) would have been prevented. The simple average of benefits

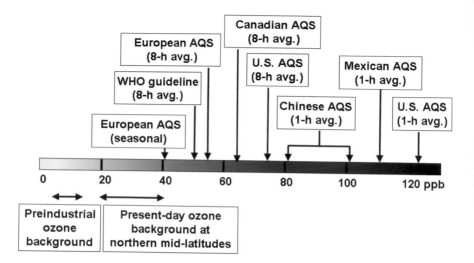

FIGURE 2.1 Ozone Air Quality Standards (AQS) in ppb (nanomoles of O_3 per mole of dry air). Different national and international standards are noted as well as estimates for northern midlatitudes of the preindustrial background (i.e., O_3 abundances with all anthropogenic emissions of NO_x, CO, VOC, and CH_4 cut off, and before current climate and stratospheric O_3 change) and the present-day baseline abundances (i.e., the statistically defined lowest abundances of O_3 in air flowing into the continents, typical of clean-air, remote marine sites). The pre-2008 U.S. AQS was 0.08 ppm, which through numerical roundoff meant that an AQS violation was 85 ppb or greater (D.J. Jacob, personal communication, 2009).

(including premature mortality) across the three years was estimated to be $4.9-$5.7 billion. If the analysis used the highest annual maximum 8-hr concentration, impacts would have increased by a factor of two to three. While highly uncertain (see Hubbell et al., 2005 for details), such estimates give a sense of the magnitude of impacts associated with ozone pollution.

In most metropolitan airsheds O_3 has a daily cycle peaking in the late afternoon. It also varies with weather patterns and over seasons. An example of the variability of surface O_3 abundance is shown is Figure 2.2 for a site in Michigan. This result is typical (i.e., violation of the U.S. NAAQS occurs episodically throughout the summer months) often for a few days in a row.

Atmospheric Chemistry and Dynamics   Ozone is a highly reactive gas that is constantly being produced and destroyed by the natural cycles of atmospheric chemistry throughout the troposphere and stratosphere. About 90 percent of the column of O_3, which protects living organisms from

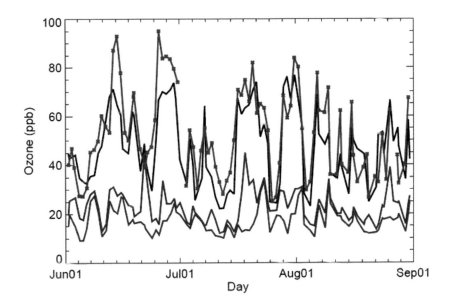

FIGURE 2.2 Time series of the daily, 8-hr, maximum O_3 abundance (ppb) for June–August 2001 at Unionville, Michigan. Observations (magenta line) are compared with model results (black line). The model attribution of O_3 sources shows the U.S. EPA policy relevant background (PRB, lowest red line) with all North American emissions cut off, and the background (blue line) with only U.S. emissions cut off. These calculations are short term and do not include the impact of CH_4 emissions on O_3, nor the impact of O_3 precursor emissions on CH_4.

SOURCE: Reprinted from Atmospheric Environment Vol. 43, H. Wang et al., Surface ozone background in the United States: Canadian and Mexican pollution influences, 2009, with permission from Elsevier.

damaging solar ultra-violet (UV) radiation, lies in the stratosphere generally more than 14 km above the surface. Even the worst pollution episodes with high O_3 abundances near the surface have little impact on the total O_3 column and do little to protect us from solar UV.

Tropospheric O_3 is a secondary pollutant; it is not emitted directly but is instead produced photochemically in the troposphere by its precursors: reactive nitrogen oxides (NO_x), carbon monoxide (CO), methane (CH_4), and other volatile organic compounds (VOCs). (Secondary organic aerosols PM is also produced under these conditions; see Chapter 3.) These precursors exist naturally but are also emitted by human activities ranging from fossil fuel use to agriculture. Open burning of biomass, either natural or caused by humans, is a large source of O_3 precursors. Another natural source of tropospheric O_3 is stratosphere-troposphere exchange. In such cases intrusions of stratospheric air (with O_3 abundances initially in excess of 100 ppb in the free troposphere, before being diluted) can contribute tens of ppb to the baseline ozone levels at the surface (Wild et al., 2003).

In addition to emission of precursors, O_3 production and loss is controlled by meteorological factors such as winds and convection that distribute precursors, clouds that control photochemistry, lightning generated NO_x, and the temperatures and water vapor that control key chemical reactions. The lifetime of an O_3 molecule in the troposphere varies according to the photochemical activity. Stevenson et al. (2006) estimate a mean tropospheric lifetime of 22 days, and Hsu and Prather (2009) derive a value of 30-35 days.

To determine the cause of elevated O_3 abundances, atmospheric chemical transport models (CTMs) allow us to separate the O_3 precursor sources with numerical experiments that turn on and off the different sources. In the case study shown in Figure 2.2, Wang et al. (2009a) calculate the different sources of surface O_3 at the Unionville, Michigan site. A slowly varying baseline level of about 20 ppb is present when all North American (NA) emissions of NO_x, CO, and VOC (but not CH_4) are stopped (lowest red line). When emissions from Canada and Mexico are added (next lowest blue line), the baseline levels are increased by a few ppb, along with some much higher episodes (+ 20 ppb) presumably due to the nearby Canadian emissions. With the addition of U.S. emissions (black line), the model reproduces much of the observed episodic behavior, including the NAAQS violations, which in some cases are clearly enhanced by non-U.S. emissions.

Defining Background and Baseline Ozone   The ultimate background level for tropospheric ozone existed before humans began to alter the atmosphere. In this preindustrial atmosphere tropospheric O_3 sources included stratospheric intrusions; lightning NO_x; natural surface emissions of CH_4, CO, and VOCs; and wildfires. This preindustrial O_3 can be defined but is not well determined by measurements (see below) or models. In EPA terms, the lowest curve in Figure 2.2 is defined as the "policy relevant background" (PRB) (i.e., without NA emissions of the short-lived precursors but with those of CH_4). It is unclear whether the PRB always includes the increase in tropospheric O_3 coming from NA sources that have already traveled around the globe, as it was in the Wang et al. global calculation cited above, since PRB can be calculated with nonglobal models. Thus, the term "background O_3" is ambiguous and will always be used here with a qualifier. The background as defined here is hypothetical and not directly measurable.

Baseline O_3 is used here (as baseline PM is used in Chapter 3) to describe a measurable quantity, the statistically defined lowest abundances of O_3 in the air flowing into a country, which is typical of clean-air, remote marine sites at the same latitude. Baseline air thus includes upwind pollution that contributes to the diffuse, uniform increase in O_3 but not the episodic events. Baseline O_3 will vary with location and season in the Northern Hemisphere and it can change over time. For example, the global emissions of CH_4 over the last several decades are responsible for the increased CH_4 abundances relative to preindustrial levels and have caused an increase in baseline O_3 of about 5 ppb (Fiore et al., 2008). Emissions of NO_x, CO, and VOC from neighboring states or the other side of the continent will also contribute to an increasingly diffuse background that is indistinguishable from baseline O_3 in most urban airsheds and beyond local control. Defining the baseline level of O_3 entering an Air Quality Management District, and how it might change, is of critical importance as it defines a minimum O_3 exposure before local pollution is added. A key point to recognize, however, is that production of O_3 is not necessarily linear, and hence the peak O_3 during a NAAQS violation does not simply shift linearly with the baseline level.

Preindustrial Ozone   The change in tropospheric O_3 since the preindustrial era is difficult to evaluate. Surface measurements at several sites in both hemispheres from the 19th and early 20th centuries require careful evaluation and calibration of the early instrumentation (Kley et al., 1988; Volz and Kley, 1988; Marenco et al., 1994; Harris et al., 1997; Staehelin et al., 1998; Pavelin et al., 1999). The best estimates are extremely low O_3 abundances (on the order of 10 ppb), which cannot be explained with current models by merely removing anthropogenic emissions of NO_x, CO, VOC, and CH_4 but rather, require large reductions in the natural sources of O_3 (Mickley et al., 2001). Multimodel studies, including most global CTMs (Prather, 2001; Gauss et al., 2006), give a range of results when the known anthropogenic emissions of O_3 precursors are removed, but they converge on a best estimate of about 33 percent increase in global tropospheric O_3 over preindustrial levels, with a greater fractional increase in the Northern Hemisphere (40-100 percent). There remains uncertainty in this value, yet when we combine the limited historical observations with our proven chemical knowledge of the formation of O_3 and the model studies, we have high confidence that current baseline levels of tropospheric O_3 in the Northern Hemisphere are much greater than the natural preindustrial levels. This elevated baseline is maintained by emissions of precursors worldwide.

Ozone the Greenhouse Gas   In terms of driving 20th-century climate change, tropospheric O_3 is the third most important anthropogenic greenhouse gas after carbon dioxide (CO_2) and CH_4. Best current estimates give a radiative forcing (year 2005 minus year 1750) of + 0.35 W m^{-2}, which can be compared with + 1.66 W m^{-2} for CO_2 (see Gauss et al., 2006; Forster et al., 2007b). Ozone in the free troposphere, not in the boundary layer over polluted regions, is responsible for most of this forcing, yet the O_3 abundances throughout the troposphere are increased by these urban and industrial emissions. Anthropogenic radiative forcing for both tropospheric O_3 and aerosols occurs primarily in the Northern Hemisphere, and there remains some uncertainty (e.g., Mickley et al., 2004; Shindell et al., 2006) as to whether these changes have disproportionate climate impact on regional scales in the northern midlatitudes.

> Finding. Combining the evidence from observations and models, and including our basic knowledge of atmospheric chemistry, there is high confidence that human activities have raised the baseline levels of tropospheric O_3 in the Northern Hemisphere by 40-100 percent above preindustrial levels. Much of this increase can be directly attributed to anthropogenic emissions of ozone precursor species (CH_4, NO_x, CO, VOCs).

> Finding. U.S. NAAQS violations (e.g., 8-hr average greater than 75 ppb) are caused primarily by a combination of regional emissions and unfavorable meteorology. These are augmented by a changing baseline and episodic events caused, for example, by wildfires, lightning NO_x, occasional stratospheric intrusions, as well as distant anthropogenic emissions. Most violations are only a few ppb above the standard, and thus the increase in baseline O_3 since the preindustrial era driven by global pollution has contributed to these violations.

CHANGING BASELINE O_3 AND IMPACT ON AIR QUALITY STANDARDS

Detecting trends in baseline O_3 is complicated by the natural variability in tropospheric O_3 due to weather patterns, seasons, and short-term climate variations. Regular measurements need to be maintained for a decade or longer, thus intensive aircraft field campaign data are not directly useful for trend analyses. Surface O_3 is routinely measured at many locations using continuous UV absorption instrumentation, a well-documented method. Vertical profiles of O_3 in the free troposphere are measured at many locations by balloon-borne sensors (ozonesondes), typically launched weekly as part of the World Meteorological Organization's Global Atmospheric Watch (WMO GAW) program. The ozonesonde data have been analyzed for trends (e.g., Oltmans et al., 2008) although the lower sampling frequency of the sondes, together with difficulties related to long-term calibration, limit the statistical detection of trends using these data (Deshler et al., 2008). For the last 15 years vertical O_3 profile measurements have been made on board civil aircraft as part of the MOZAIC program (Marenco et al., 1998), which provide the more frequent measurements needed for trend detection at certain airports.

Given that tropospheric O_3 increased sometime over the 20th century due to increasing emissions of precursors including CH_4, the present rate of increase is a key uncertainty. Comprehensive reviews of Northern Hemisphere O_3 trends have been presented by Vingarzan (2004), Cape (2008) and Royal Society (2008). Ozone trends vary substantially from place to place. The majority of sites with long-term observations indicate an increasing trend in O_3 over the last two to three decades. The trends are in the range of 0.5-2 percent/yr (approximately 0.2-0.8 ppb/yr). Over the most recent decade the positive trend appears to have leveled off at some sites (e.g., Oltmans et al., 2006), while at others, such as the Atlantic coastal site at Mace Head, Ireland, and numerous rural sites in the western United States, O_3 continues to increase. A statistically robust trend in baseline O_3 of 0.2-0.3 ppb/yr has been reported at Mace Head for the period of 1987-2007 (Derwent et al., 2007). The baseline O_3 abundances at Mace Head (e.g., 55 ppb in March 1999) are high compared with similar sites on the U.S. west coast.

Along the west coast of the United States a number of remote sites have been used to derive trends in O_3 (Jaffe et al., 2003; Parrish et al., 2004b, 2008). Farther inland, Jaffe and Ray (2007) analyzed data from rural CASTNET sites. While a variety of methods have been used by the different groups, the results are consistent in suggesting an increase in baseline O_3 of 0.3-0.4 ppb/yr. These trends are present in all seasons, and Figure 2.3 shows data from eight rural CASTNET sites in the western U.S. (Jaffe and Ray, 2007). These results are consistent with earlier work that suggests an increasing contribution to surface O_3 in the United States from external sources during spring (e.g., Lin et al., 2000) and possibly other seasons as well. In England an average trend of 0.14 ppb/yr was reported for 13 rural sites over the period 1991–2006 (Jenkin, 2008).

In the free troposphere O_3 data records are much more limited. Analysis by Logan et al. (1999) and Prather Figure 4.8 (2001) shows an unsteady but clear increase in O_3 in the middle troposphere of the northern mid-latitudes of about 0.2 ppb/yr for the period 1970–1997. At several mountain stations in Europe a positive trend in O_3 of 0.4-0.5 ppb/yr has been reported for 1991–2002, with the most statistically robust trends reported for fall, winter, and spring (Royal Society, 2008). Using a 10-year record of ozonesondes from a U.S. west coast location Oltmans et al. (2008) reported no significant trends. Oltmans et al. (2006) report significant regional variations in the trends, compared the more consistent patterns seen in

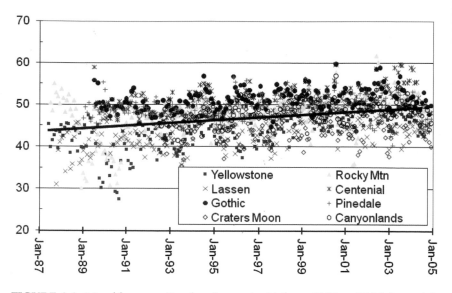

FIGURE 2.3 Monthly mean O_3 abundances (ppb) from 1987 to 2005 from eight rural CASTNET sites in the western U.S. The average seasonal pattern at each site has been removed and a linear regression fit to the data: +0.34 ppb/yr.
SOURCE: Reprinted from Atmospheric Environment Vol. 41, D. Jaffe and J. Ray, Increase in surface ozone at rural sites in the western U.S., 2007, with permission from Elsevier.

surface observations. For example, they report significant increases in tropospheric O_3 over Europe and Japan in the 1970s and 1980s, but a much smaller increase over the United States. Tanimoto et al. (2008) identified a significant increasing trend in free tropospheric O_3 downwind of China for 1998-2006, a period of rapid growth in Chinese NO_x emissions. Likewise, Bortz et al. (2006) report a large trend, 1.1 ppb/yr, in the northern tropics for the MOZAIC flight-level data for 1994-2003. While evidence generally shows that more regions of the Northern Hemisphere have increasing O_3 baselines than have constant or decreasing ones, the magnitudes and specific patterns are not fully understood or simulated in chemical-transport models. Changes in natural variability, which can influence, for example, the stratospheric flux of O_3 into the troposphere, could be a contributing factor.

Impact on Air Quality Standards   Long-range transport of O_3 in pollution plumes can be seen as distinct episodes of elevated O_3 (e.g., Figure 1.3). Plumes of Asian origin with O_3 abundances exceeding 80 ppb are observed from aircraft in the free troposphere over the United States (Nowak et al., 2004; Liang et al., 2007). Observations at remote surface sites do not show detectable long-range pollution events for ozone, and this likely reflects the dilution during entrainment into the boundary layer (Hudman et al., 2004). Direct boundary layer transport from Asia is not a major contributor to the North American baseline because of the unfavorable circulation and the fast chemical loss of ozone under moist, sunlit conditions. Such transport does occur between North America and Europe and may raise the baseline O_3 observed coming into Europe.

Models and observations do, however, indicate that the intercontinental pollution influence on U.S. surface O_3 manifests itself as a large-scale, diffuse increase in baseline O_3 which is due to O_3 and its precursors being transported around the globe by fast winds in the free troposphere and then subsiding and mixing into the surface layer. One estimate of the increase in U.S. baseline O_3 due to all non-U.S. emissions is 5-10 ppb with relatively little variability (Fiore et al., 2003a). This baseline enhancement can be characterized as an increase in the frequency of AQS violations (Fiore et al., 2002), but this may be misleading in that violations determined by a threshold can be caused by any small contribution to the total. A better measure of the ability of Air Quality Management Districts to control their AQS violations is the relative importance of domestic vs. distant emissions in contributing to local surface O_3 (Jacob et al., 1999; Fiore et al., 2008).

In 2008 the U.S. NAAQS O_3 standard was tightened from an 8-hr average of 0.08 ppm (effectively 85 ppb) to 75 ppb, thus increasing the importance of the baseline O_3 abundances controlled by global emissions of O_3 precursors. The O_3 standard is evaluated relative to the PRB, which is presumably not amenable to North American regulation. The argument that tighter ozone standards are unachievable is based on some observations of O_3 abundances at remote sites that are often in excess of 60 ppb (Lefohn et al., 2001), and presumably represent the PRB. Recent modeling and analysis refutes this, finding a 20-40 ppb PRB for the United States and noting that those larger abundances are associated either with high-elevation sites or with more distant North American pollution (Fiore et al., 2003b).

Finding. Baseline tropospheric O_3 abundances at many remote locations in the Northern Hemisphere have changed over the last few decades at rates ranging from hardly at all to as much as 1 ppb/yr. The causes of these changes are not clear.

Recommendation. The measurements documenting changes in baseline O_3 over the last few decades need to be systematically and collectively reviewed using consistent methods of analysis. Observations of O_3 and related trace species should be placed in perspective through global chemical-transport modeling that includes the history of key factors controlling tropospheric O_3: anthropogenic and natural emissions of precursors, stratospheric O_3, and climate. An analysis of the key uncertainties should be undertaken to ensure that future changes can be attributed with greater confidence.

DIRECT OBSERVATION OF LONG-RANGE TRANSPORT OF O_3 AND PRECURSORS

A variety of observational tools are used to identify long-range transport of O_3 and its precursors. Satellite datasets provide a continuous large-scale view and can sometimes track large plumes originating from intense pollution events as they cross the hemisphere (Edwards et al., 2004; McMillan et al., 2008; Zhang et al., 2008a). These observations are generally limited, however, to very intense pollution plumes and a few species. A spaceborne lidar (CALIPSO) has recently provided useful information on vertical structure by measuring the aerosols in plumes. As noted above, ozonesondes, the MOZAIC data (Thouret et al., 2006), and ground-based O_3 lidars (Colette and Ancellet, 2005), also provide high-resolution vertical profiles that can be used to identify pollution plumes, but not map their extent or transport in the same way as satellites. Brief, intensive field campaigns involving a range of in situ measurements and mobile lidars are often able to map out tropospheric ozone and provide important detail on tropospheric O_3 plumes, including precursors, vertical profiles, and their dilution into the baseline levels. Figure 2.4 shows an aircraft-measured curtain of tropospheric O_3 on a transect from Japan to Hawaii during TRACE-P (Wild et al., 2004b). The corresponding CTM simulation is able to broadly match the pollution plumes from Asia, the stratospheric intrusions, and the sharp gradients across a frontal system. Such intensive field campaigns involving coordinated flights from several locations have been successful in providing data on transport pathways and provide an integrating step between the measurements from satellite and distant ground-based monitoring sites located in source and receptor regions.

Long-range tropospheric ozone transport can potentially be detected by satellite observation of tropospheric O_3 columns that are derived using residual techniques (i.e., subtract a stratospheric component, which accounts for about 90 percent of the total, from the total column measurement with the aid of other measurement or model information). Unfortunately, there is often a strong coincidence of extra-tropical tropospheric ozone column anomalies with probable troposphere-stratosphere exchange events or folds (Schoeberl et al., 2007; Zhang et al., 2008a), and these may be mistaken for pollution events since folds often occur in the Atlantic and Pacific pollu-

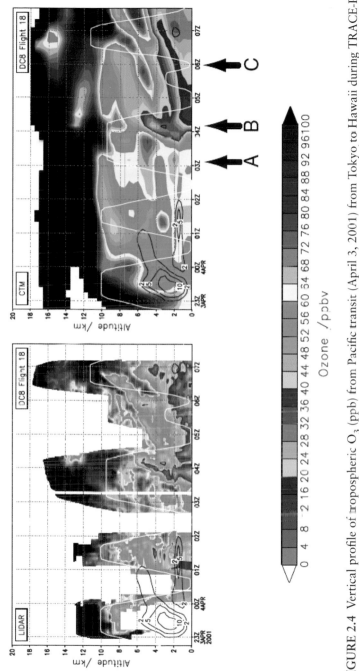

FIGURE 2.4 Vertical profile of tropospheric O₃ (ppb) from Pacific transit (April 3, 2001) from Tokyo to Hawaii during TRACE-P. The DC-8 lidar observations on the left show (A) an O₃ pollution plume from Asia (02Z-04Z, 1-6 km), (3) a strong frontal passage (04Z, 0-4 km), and (C) stratospheric intrusions (05Z-07Z, 4-9 km) that are reasonably well simulated by the model results, shown on the right (Wild et al., 2004b).

tion corridors. Nevertheless, monthly averaged tropospheric column ozone (TCO) derived from the Aura satellite (MLS and OMI) indicate significant O_3 enhancements in the regions of greatest precursor emission (i.e. eastern U.S., Europe, East Asia, and a broad enhanced region in the midlatitudes associated with pollution export from the continents (Figure 2.5) (Ziemke et al., 2006). Seasonally, the TCO follows the Northern Hemisphere industrial pollution in the July and the Southern Hemisphere biomass burning source of O_3 in November.

CO is the only long-lived O_3 precursor readily measured from satellite that can provide a good indication of long-range transport. Intense pollution sources can produce plumes with enhancements of > 100 percent over background values. See Figure 2.6 for an example of CO pollution plumes observed with the Terra/MOPITT satellite instrument. Correlations of enhanced O_3 and CO within pollution plumes measured by Aura/TES suggest initial export of O_3 in the plumes from Asia along with continued production of O_3 during transpacific transport (Zhang et al., 2008a). These correlations are also used to analyze in situ measurements to identify sources of plumes (forest fires vs. fossil fuel combustion) and to try to track them during long-range transport (Parrish et al., 1993; Stohl et al., 2002).

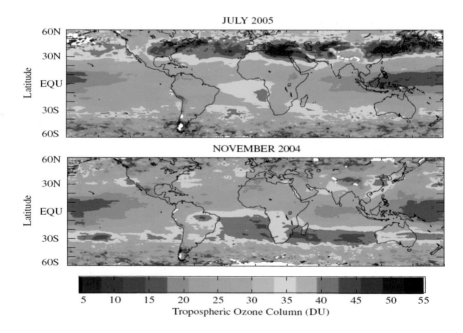

FIGURE 2.5 Monthly averaged OMI/MLS TCO (in Dobson units) for July 2005 and November 2004 (Ziemke et al., 2006).

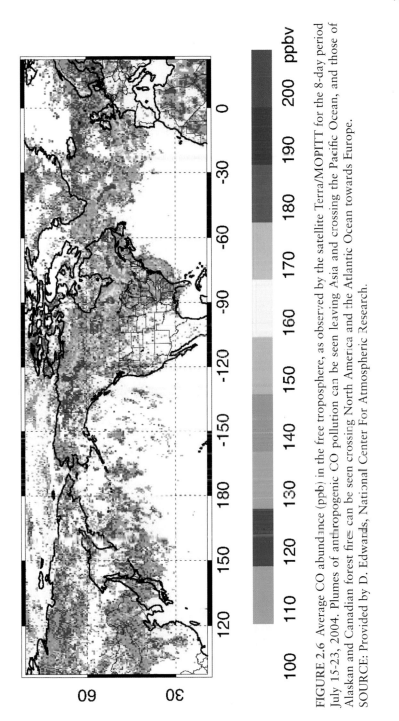

FIGURE 2.6 Average CO abundance (ppb) in the free troposphere, as observed by the satellite Terra/MOPITT for the 8-day period July 15-23, 2004. Plumes of anthropogenic CO pollution can be seen leaving Asia and crossing the Pacific Ocean, and those of Alaskan and Canadian forest fires can be seen crossing North America and the Atlantic Ocean towards Europe. SOURCE: Provided by D. Edwards, National Center For Atmospheric Research.

Such correlations are used to deduce photochemical production and loss rates for CO and O_3 during transit, but other factors such as the large-scale gradients along the plume and in background abundances complicate these analyses (Chin et al., 1994; Pfister et al., 2006; Real et al., 2008).

The availability of nearly a decade of satellite CO observations has led to the derivation of multiyear regional CO inventories through inverse modeling (Pétron et al., 2004; Kopacz et al., 2009) that has increased our emission estimates for CO from fossil and biofuel use in Asia by a large factor, and improved our understanding of the emission patterns. CO can also be used as a proxy for inferring emissions and distributions of other species that are readily measured to correlate with CO using in situ measurements but are not readily measured directly from satellite. Although satellite measurement of nitrogen dioxide (NO_2) and formaldehyde (HCHO) clearly indicate source regions for these O_3 precursors and provide emissions estimates (Fu et al., 2007; Wang, 2007), long-range transport of these species is not expected nor readily observed because of their short lifetimes (Martin et al., 2006).

Aircraft measurement campaigns feature a diversity of well-calibrated instruments that can measure multiple species and allow diagnosis of the chemical and dynamical processes influencing O_3 during long-range transport, providing a critical test of chemistry and transport models. A recent example is the ICARTT campaign that included components from North America (INTEX-NA, NEAQS) and Europe (ITOP), as well as a Lagrangian experiment (IGAC Lagrangian 2K4) with dedicated flights using multiple aircraft to sample the same air masses several times as they crossed the North Atlantic between North America and Europe. Analysis of an anthropogenic plume transported at low levels and a forest fire plume that descended into Europe showed O_3 destruction in the former and production in the latter (Real et al., 2007).

Spring import of enhanced O_3 levels into western North America has been identified in analysis of surface data and aircraft measurements during various campaigns focused on quantifying transport of polluted air masses from Asia to North America (e.g. TRACE-P, INTEX-B, ITCT-2K2). For example, analysis of INTEX-B data suggests that Asian pollution plumes contributed 9 ± 3 ppb to an O_3 abundance of 54 ppb at Mount Bachelor, Oregon (2.7 km altitude) during April-May 2006 (Zhang et al., 2008a). Results based on the GEOS-CHEM model suggest that about half of the pollution O_3 was produced in the emission region and half during transport across the Pacific, largely as a result of the decomposition of PAN (peroxyacetyl nitrate, an urban air pollutant formed in the air as part of photochemical smog) (Heald et al., 2003). Long-range transport of pollutants from Europe to North America is expected but remains much less well quantified.

Transoceanic transport of pollution O_3 takes place mostly in the free troposphere where meteorological conditions occasionally favor the preservation of distinct plumes as compared with the marine boundary layer. In addition, O_3 has a longer atmospheric lifetime in the free troposphere than in the warmer boundary layer (Heald et al., 2003; Price et al., 2004), thus allowing individual Asian plumes to be observed at mountain sites in the western United States (Jaffe et al., 2005b; Reidmiller et al., 2009b). Corresponding plumes are not generally observed at the surface (Goldstein et al., 2004; Hudman et al., 2004), presumably because of the dilution into large air masses in the boundary layer, or because of loss during boundary layer transport. Substantial production of O_3 in boreal forest fire plumes may also lead to pollution events in downwind regions (Real et al., 2007). Better understanding and modeling of the entrainment of freetropospheric O_3 plumes into the boundary layer in the impacted region is needed to improve our estimates of the impact of long-range transport of pollution O_3 on AQS.

Finding. Plumes containing high levels of O_3 and its precursors (NO_x, CO, VOCs) can be transported between continents and are observed downwind of the major industrial regions and large wildfires in North America, Europe, and Asia. They are observed in the free troposphere over impacted regions but rarely at the surface due to dilution in the boundary layer.

Recommendation. Conduct focused research efforts that couple measurements with models to quantify the process of air exchange between boundary layer and free troposphere, in order to better understand how free tropospheric O_3 enhanced by long-range transport is mixed to surface.

MODELING AND ATTRIBUTION OF O_3 FROM GLOBAL SOURCES

HTAP and Other Model Results   In this section we focus our attention primarily on estimates of ozone import and export into and out of North America (NA). Attribution of the amount of O_3 produced from emissions from different regions or sources can be derived either by keeping track or tagging the O_3 formed over or downwind from the particular region (e.g., Derwent et al., 2004; Sudo and Akimoto, 2007) or by reducing or increasing particular emissions over the region of interest (e.g., Yienger et al., 2000; Wild and Akimoto, 2001; Auvray and Bey, 2005). Chemistry-transport models (CTMs) or chemistry-climate models (CCMs) are also used to perform budget studies that estimate the contributions from different sources over a particular region. For example, Pfister et al. (2008) used the global MOZART CTM to estimate that the fraction of summertime

U.S. O_3 in 2004 originating from stratosphere (26 ± 6 percent) was comparable to that from U.S. emissions (25 ± 9 percent) with smaller contributions from Eurasian sources (13 ± 5 percent), lightning (10 ± 2 percent), and boreal forest fires (3 ± 2 percent) that were very active in Alaska during this particular year. Other studies also highlight the importance of lightning NO_x over North America (Hudman et al., 2006; Singh et al., 2007). These results illustrate the complexity of the natural and anthropogenic sources influencing ozone distributions over the United States.

As part of HTAP-TF (2007) a thorough analysis of results from many different (ensemble) model simulations has been performed with the aim to reduce uncertainties by combining results from models with different representations of emissions, transport, and chemical processes. A consistent set of analysis metrics was also used (see http://htap.icg.fz-juelich. de/data/FrontPage for details). Of particular relevance to this report are results from 21 global and hemispheric CTMs that were used to estimate the change in surface ozone resulting from reduced emissions over East Asia (EA), North America (NA), Europe (EU), and South Asia (SA) (Fiore et al., 2009). Note that NA, EA and EU regions have slightly different areas and SA is a factor of two lower. About half the models had a horizontal resolution of 3×3 degrees or finer. For each region, anthropogenic emissions of NO_x, CO, and VOCs from all sectors, including shipping, were reduced by 20 percent (first for each individual sector, then for all of them together) for model simulations of the year 2001. The 20 percent changes in ozone precursor emissions used in HTAP are comparable with recent changes: U.S. total NO_x emissions decreased about 3 percent/yr between 2000 and 2006, mostly in the eastern U.S. (EPA-TTN; http://www.epa.gov/ttnchie1/trends/); whereas East Asian emissions increased approximately 7 percent/yr for the years 2000-2005 (Ohara et al., 2007).

Responses to reductions in foreign emissions (not including methane) were found to be larger in spring and fall, with largest responses to reductions in NO_x emissions. Import sensitivities (defined as the ratio in surface O_3 response to a 20 percent decrease in the three foreign source regions vs. a 20 percent decrease in the domestic emissions) peak during the winter and spring or late fall (0.5 to 1.1) when the response of O_3 to domestic emissions is small (Figure 7 in Fiore et al., 2009). Note that "domestic" as used here does not mean local emissions but includes all interstate and even international emissions within the NA region. During the summer months when domestic O_3 production is at a maximum, import sensitivities for all regions are small (0.2-0.3). The number of days with 8-hr average O_3 abundances greater than 60 ppb are much more sensitive to reduction in domestic emissions than in foreign (Reidmiller et al., 2009a).

Figure 2.7 shows the modeled change in 24-hr average surface O_3 over NA resulting from a 20 percent change in anthropogenic precursor

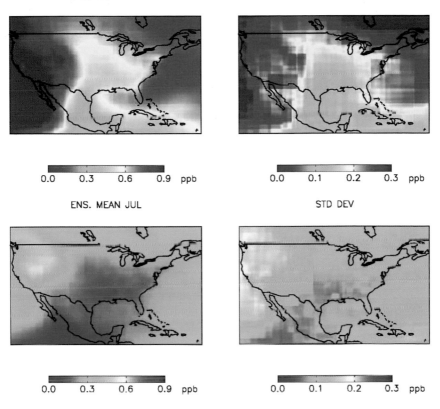

FIGURE 2.7 Model mean surface O_3 (ppb) for April and July over North America that is attributable to 20 percent of the anthropogenic emissions of O_3 precursors (NOx, CO, VOC) from the other three major industrial regions of the NH (EA, EU, SA). The standard deviation of the ensemble of 14 HTAP models is also shown (Fiore, personal communiation. See Fiore et al., 2009 for details).

emissions (NO_x, CO, VOC) from the combined foreign source regions (EA + SA + EU), where the O_3 change results from emissions being adjusted from 80 to 100 percent of current levels. The April and July monthly means from the ensemble of 14 HTAP models along with the standard deviation of the ensemble are plotted. Peak levels of imported O_3 can be seen over the west coast of North America, with higher values in spring than in summer. Smaller increases are also seen over eastern Canada and off the U.S. east coast. Interestingly, there is also predicted import into Florida in November, arising from transport of emissions from Europe around the Azores High during the winter months. EU and EA sources (not shown separately) are dominant, contributing about equally (Figure 5a in Fiore et al., 2009).

The HTAP analysis also considered the impact of NA on other regions. For example, while EA is mainly influenced by EU emissions (particularly in spring), there is also a non-negligible contribution from NA emissions. Their results also confirmed previous studies showing that NA is the main external contributor to surface EU O_3 due to its relatively close proximity. In those studies (summarized in Table 5.2 in HTAP-TF, 2007) ozone enhancements varied between 1 to 5 ppb but sometimes as high as 10 ppb for particular events.

The recent HTAP results based on ensemble model results show less variability and lower enhancements since these estimates provide an estimation of average increases in baseline O_3 due to different source regions and not due to particular events. Derwent et al. (2008) also examined regional influences in source-receptor relationships affecting European O_3 by conducting a series of increased NO_x pulse experiments over 10×10 degree grids covering East Asia and North America. Largest responses to NA emissions were found at sites in Western Europe. In winter modeled O_3 responses at Mace Head were more sensitive to emissions in northern NA whereas in summer emissions from eastern NA produced higher responses. Since O_3 has a shorter lifetime in summer, closer pollution sources have a greater influence. This location showed positive responses in the short term (up to several months) followed by smaller negative responses in the longer term (1 year or more) due to feedbacks between NO_x, OH, O_3, and CH_4 chemistry. This resulted at certain locations in an integrated net negative response, especially in summer (e.g., to changes in southern U.S. emissions).

Changing NO_x, CO, or VOC perturbs OH and thus the CH_4 lifetime that in turn changes CH_4 and baseline O_3 abundances over decades (Wild et al., 2001). Fiore et al. (2009) also tried to quantify some of these short- versus long-term feedbacks in O_3 responses; reducing NO_x emissions over all regions lowers OH and increases CH_4, leading to reductions in the short-term decrease in surface O_3 by 15-20 percent; reducing CO or non- methane VOCs by 20 percent has the opposite effect and leads to additional increases in surface O_3 of around 35 percent and 10 percent, respectively. Combined emission reductions offset each other leading to long-term effects that were always less than 3 percent.

This concept of different timescales influencing the impact of O_3 responses in downwind receptor regions has also been examined using the adjoint modeling approach. For example, Figure 2.8 shows the sensitivity of the mean surface O_3 concentrations at Trinidad Head, California, to production over upwind source regions (Hakami et al., 2007). Local O_3 sources have the greatest impact, although the Asian contribution from eastern China and Japan and O_3 produced during transport over the North

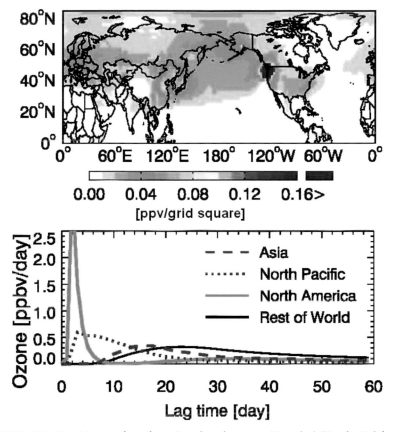

FIGURE 2.8 Sensitivity of surface O_3 abundance at Trinidad Head, California, to ozone production in different regions of the world, as inferred from the GEOS-Chem adjoint model for the period April 17 - May 15, 2006 (Zhang et al., 2009a). Left: sensitivities integrated in time over the depth of the tropospheric column. Right: time-dependent sensitivities (going back in time) to O_3 production over Asia, the North Pacific, North America, and the Rest of World. These results help demonstrate that production over any one region (e.g., North America) can include long-range transport of distant sources of O_3 precursors as well as local emissions (Kotchenruther et al., 2001).

Pacific plume are significant. The time lag in the response (impact) at Trinidad Head is about 16 days.

Emissions of O_3 precursors in Asia have been increasing rapidly (Irie et al., 2005; Richter et al., 2005; Ohara et al., 2007), and this has resulted in increased production and outflow of O_3, especially during spring (Tanimoto et al., 2008; Xu et al., 2008; Zhang and Tao, 2009). As a result some researchers have suggested that the increasing Asian emissions are responsible for the trends in spring O_3 along the west coast of the United States (Jaffe et al., 2003b; Parrish et al., 2004). The HTAP assessment (Fiore et al., 2009; Reidmiller et al., 2009a) can be used to evaluate the significance of increasing Asian emissions and the consistency with observations. According to Reidmiller et al. (2009b) a 20 percent reduction in East Asian emissions would reduce the spring average maximum daily 8-hr average in the western United States by approximately 0.4 ppb. Assuming linearity, a 10 percent/yr increase in Asian precursor emissions would correspond to an increase in MDA8 of ~ 0.2 ppb/yr, which is similar in magnitude to the 0.34 ppb/yr increase in mean daytime O_3 reported by Jaffe and Ray (2007). Although the observed increase in O_3 was found in all seasons, the HTAP assessment suggests that only during winter-spring is the increase consistent with the increasing Asian emissions. During summer, an increase in wildfires across the western United States is the most likely cause for the increase in O_3 (Jaffe et al., 2008).

Model Uncertainties   The results presented in the previous sections rely largely on simulations using global models. Even though results from many different models were used in the HTAP analysis, there are still significant uncertainties in individual model calculations. Comparison with observations is used as a means of evaluating model performance. For example, Fiore et al. (2009) attempted to assess model skill by comparing the HTAP ensemble model mean with surface O_3 observations in the United States, Europe, and Asia. They found that modeled monthly mean surface O_3 concentrations capture observed seasonal cycles at European sites reasonably well, but overestimated summertime concentrations over Japan and the eastern United States (more than 14 ppb in July). Biases were particularly large over the southeastern United States and may be related to uncertainties in the chemical reactions among isoprene, NO_x, and O_3. A similar but more detailed analysis focusing on the United States also reports a significant positive bias in the HTAP model ensemble results for summer in the eastern United States, along with a negative bias of surface O_3 from the ensemble mean for the western United States during spring (Reidmiller et al., 2009a). Large uncertainties in the emissions may be significant sources of error in these model simulations. For example, in the HTAP exercise, emissions of nonmethane VOCs varied by a factor of 10 between models.

   To evaluate long-range transport of pollutants, more rigorous validation of models is required where model performance is evaluated in terms of their ability to simulate pollutant plume transport and dilution (to baseline) and chemical and physical processing,including transport of plumes into downwind receptor regions where mixing into the boundary layer is important. Comparisons with data collected during dedicated aircraft campaigns are able to pinpoint particular errors in global models related to emissions, dynamics, or wet deposition and thereby reduce uncertainties. For example, Hudman et al. (2008) used INTEX-NA data to conclude that EPA CO emissions are too high by 60 percent. In another study based on analysis of INTEX-B data, Zhang et al. (2008a) added weight to previous work (Palmer et al., 2003) suggesting that Asian emissions are too low by a factor of two. Analysis of ICARTT data collected in the same air mass during transport across the North Atlantic also showed that global models are unable to capture the evolution of, for example, forest fire plumes related to insufficient resolution or missing or inaccurate photochemistry (e.g., Real et al., 2007).

Global models need to be run at higher horizontal and vertical resolutions and combined with Lagrangian techniques or regional air quality models in order to capture plume transport between continents. The rate at which pollutant plumes mix with so-called background air determines whether a plume event will be detected over a receptor region or whether the signal will contribute to enhanced baseline concentrations. Recently, Pisso et al. (2009) estimated that global models need to be run with at least 40 km horizontal and 500 m vertical resolution in order to simulate long-range transport of pollutant plumes. The impact of resolution errors on modeled O_3 production rates is less clear. While Wild and Prather (2000) found linearly decreasing errors ranging from 27 percent at 5.5 degrees to 5 percent at 1.1 degrees, Esler et al. (2004) suggested that errors might become nonlinear below 1 degree and possibly much higher (20-50 percent) in the sharp gradients at plume edges.

Finding. Baseline O_3 levels in the Northern Hemisphere are elevated by current anthropogenic emissions of NO_x, CO, and VOCs. Multimodel studies calculate that a 20 percent reduction in these emissions from any three of the four major industrial regions of the Northern Hemisphere will reduce surface O_3 in the fourth region by about 1 ppb on average (i.e., a baseline enhancement) but with large spatial and seasonal variation. Results of similar magnitude are found for a 20 percent reduction in CH_4 emissions. Unfortunately, the range of results across this multimodel ensemble is comparable to the average result. These calculations were conducted for 2001 and in the following eight years, emissions have changed by 20 percent or more in some regions (e.g., decreasing in the United States, increasing in East Asia).

Recommendation. As models provide the only practical method of attributing a fraction of surface O_3 and possible AQS violations to different sources of pollution, a major effort must be made to calibrate, test, and improve these models with a wide range of atmospheric chemistry measurements ranging from individual events to climate statistics of air pollution. These tests must span the range of necessary modeling from global transport and stratosphere-troposphere exchange down to the urban airshed. While improving the models is a priority, realistically quantifying their uncertainty is key. Development of adjoint and ensemble modeling approaches will be important in assessing NAAQS compliance and intracontinental pollution under rules such as the U.S. EPA Clean Air Interstate Rule.

HEALTH IMPACTS OF IMPORTED O$_3$

Acute exposure to elevated O$_3$ levels is associated with increased hospital admissions for pneumonia, chronic obstructive pulmonary disease, asthma, allergic rhinitis and other respiratory diseases, and with premature mortality (e.g., Mudway and Kelly, 2000; Gryparis et al., 2004; Bell et al., 2005; Ito et al., 2005; Levy et al., 2005; Bell et al., 2006; NRC, 2008; Jerrett et al., 2009). A 10-ppb increase in 1-hr daily maximum ozone is associated with a 0.41-0.66 percent increase in mortality (Gryparis et al., 2004; Ito et al., 2005; Levy et al., 2005). Outdoor ozone concentrations and activity patterns are the primary determinants of ozone exposure (Suh et al., 2000; Levy et al., 2005). A recent NRC committee concluded that "the association between short-term changes in ozone concentrations and mortality is generally linear throughout most of the concentration range. . . . If there is a threshold, it is probably at a concentration below the current ambient air standard." (NRC, 2008). In addition, there is weak evidence that chronic exposure to ozone increases mortality; if confirmed, the total health burden of exposure to ozone would be much higher than current estimates (NRC, 2008).

Anenberg et al. (2009) estimated the impacts of intercontinental O$_3$ transport on mortality using the multimodel ensemble mean surface O$_3$ responses to perturbation scenarios produced by the HTAP analyses (HTAP-TF, 2007; Fiore et al., 2009) and health impact functions. A health impact function is used to calculate avoided deaths in a population, using baseline mortality rates, the modeled O$_3$ changes, and a concentration-response factor.[1] For cardiopulmonary deaths the concentration response function

---

[1] More specifically, premature mortalities avoided due to the O$_3$ concentration change are calculated as: $\Delta Mort = y_0 (1-exp^{-\beta \Delta X}) Pop$, where β is the concentration-response factor (CRF, log-linear relationship between O$_3$ concentration and relative risk, defined from the literature), ΔX is the change in O$_3$ concentration; y_0 is the baseline mortality rate (y_0) and Pop is the exposed population. This equation is applied in each grid cell for each month using the corresponding population and baseline mortality rates, and results are summed to yield annual avoided premature mortalities. CRFs are from a daily time-series study of the average relative risk of mortality associated with short-term ambient O$_3$ concentrations in 95 U.S. cities (Bell, 2004).

was based on Bell et al. (2004): a 0.64 percent increase in cardiovascular and respiratory mortality for each 10 ppb increase in 24-hr average O_3 with a 95 percent confidence interval of 0.31 to 0.98 percent. The avoided nonaccidental mortality is also evaluated and shown to be larger, since it includes cardiopulmonary mortality.

As shown in Tables 2.1 and 2.2, 20 percent reductions of anthropogenic NO_x, nonmethane VOCs, and CO emissions in NA were estimated to avoid more deaths outside NA than within (68-76 percent of resulting avoided deaths in the Northern Hemisphere [NH] occur outside NA). Reductions in EU were also estimated to avoid more deaths outside EU than within when a low-concentration threshold was applied (55-58 percent of resulting avoided deaths in the NH occurred outside of EU). The opposite was true for EA and

TABLE 2.1 Annual avoided cardiopulmonary deaths (in hundreds) in each receptor region and in the entire NH, when anthropogenic NO_x, NMVOC, and CO emissions are reduced by 20 percent within each region.

Source Region	Reception Regions				
	NA	EA	SA	EU	NH
NA	9 (4 , 13)	7 (3 , 10)	6 (3 , 9)	11 (5 , 17)	36 (18 , 55)
EA	2 (1 , 3)	43 (21 , 66)	6 (3 , 9)	5 (3 , 8)	59 (29 , 91)
SA	1 (0 , 1)	4 (2 , 6)	76 (37 , 120)	2 (1 , 3)	85 (41 , 130)
EU	2 (1 , 3)	8 (4 , 12)	6 (3 , 10)	17 (8 , 26)	38 (18 , 58)

Note: This assumes no low-concentration threshold for pollution impacts. Confidence intervals (95 percent) reflect uncertainty in the concentration response function only (Bell et al., 2004). SOURCE: Anenberg et al. (2009).

TABLE 2.2 Annual avoided nonaccidental deaths (in hundreds) in each region and in the entire Northern Hemisphere when anthropogenic NO_x, NMVOC, and CO emissions are reduced by 20 percent within each region.

Source Region	Receptor Regions				
	NA	EA	SA	EU	NH
NA	16 (1 , 30)	10 (6 , 14)	10 (6 , 15)	17 (11, 23)	60 (27 , 94)
EA	3 (2 , 5)	65 (15, 120)	10 (5 , 16)	8 (5 , 11)	93 (26 , 160)
SA	1 (0 , 2)	6 (2 , 10)	132 (80,180)	3 (1 , 5)	148 (89 , 210)
EU	3 (2 , 4)	12 (6 , 18)	11 (6 , 17)	25 (-9 , 58)	60 (9 , 110)

Note: This assumes no low-concentration threshold for pollution impacts. Confidence intervals (68 percent) are derived from ± 1 standard deviation of the model ensemble's O_3 change. They do not reflect uncertainty in the concentration response function. Note that nonaccidental deaths is a larger class that includes cardiopulmonary deaths (from Table 2.1). SOURCE: Anenberg et al. (2009).

SA, with 70 percent of the resulting NH avoided deaths occurring within the source region for EA and 90 percent for SA. Reducing emissions in any of the four regions resulted in many annual avoided deaths in EA and SA, due to large populations and high baseline mortality rates. Reducing anthropogenic CH_4 emissions in each region by 20 percent gives about half the number of avoided mortalities in the NH as does the 20 percent reduction in NO_x, CO, and VOC emissions. Reductions in CH_4, however, will also reduce O_3 and mortality in the tropics and SH (West et al, 2006).

The relative importance of source-receptor pairs for mortality was strongly influenced by the accuracy and consistency of the HTAP model ensemble, and this had disproportionately large standard deviations for O_3 responses in the same region where emissions were reduced. None of the HTAP models used to generate the O_3 response to emission reductions has the necessary resolution to simulate urban air quality; hence the domestic response is likely underestimated. The more distant response to emissions does depend on resolution (see discussion above) but less so. In general, the impact of ozone-modeling uncertainty was greater than that of concentration-response-function uncertainty.

Finding. With high confidence we can state that increases in O_3 occur in populated regions due to distant pollution and such increases are detrimental at some level to human health, agriculture, and ecosystems. A preliminary study finds that about 500 premature cardiopulmonary deaths could be avoided annually in North America by a combined 20 percent reduction in NO_x, CO, and VOC emissions from the other three major Northern Hemisphere industrial regions and, correspondingly, about 1800 in Europe. The uncertainty in these estimates is large, at least ± 50 percent, and reflects uncertainties in modeling both O_3 change and health effects.

Recommendation. Perform additional research to (1) attribute quantitatively the surface O_3 change to distant emissions; (2) specify the shape of the exposure-response curve (i.e., whether a threshold exists); (3) estimate avoided premature health burdens across a wider range of health outcomes; and (4) estimate avoided premature health burdens for particularly sensitive groups (i.e., children and others).

THE FUTURE—CHANGING CLIMATE AND EMISSIONS

Impact of Climate Change on Local Pollution Episodes Current observations and model projections show a warming climate. This climate change leads to increased atmospheric water vapor that causes substantial O_3 reductions especially in the tropical lower troposphere, and it also leads to enhanced stratosphere-troposphere exchange, which increases the flux of stratospheric O_3 into the troposphere (Collins et al., 2003; Sudo et al., 2003; Shindell et al., 2006; Zeng et al., 2008). In addition, we expect that recovery of the stratospheric ozone depletion caused by anthropogenic halocarbons will also lead to increases in the stratospheric O_3 flux (e.g., Fusco and Logan, 2003). These major climatic changes will drive baseline O_3 in different directions depending on latitude and season. For example, photochemical reduction in O_3 is greatest at low latitudes while stratospheric-driven increases will dominate high latitudes. Lightning-generated NO_x, possibly the largest natural O_3 source in the tropics, will increase if deep convection increases in a warmer climate (Toumi et al., 1996). Current model projections do not provide an adequate scientific consensus due to the net effect of 21st-century climate change.

Polluted U.S. sites show a strong correlation of high-ozone episodes with elevated temperature (Lin et al., 2001). This correlation is well reproduced in models and is driven in part by chemistry, biogenic VOC emissions, and the association of high temperatures with stagnation events that trap pollution (Jacob et al., 1993; Sillman and Samson, 1995). There appears to have been an increase in the frequency of stagnation events in the eastern United States over the past decades, compromising progress to achieve the ozone AQS (Leibensperger et al., 2008). In one model this specific pattern is predicted as a consequence of current climate change due to northward shift of storm tracks (Mickley et al., 2004). Thus, a range of evidence projects that 21st-century climate change will increase the O_3 AQS violations driven by local pollution.

Quantifying these effects requires 21st century projections using global climate models (for the O_3 baseline) as well as extrapolation of climate change to regional scales relevant to AQS with either separate regional models or statistical methods to downscale the meteorology. A number of such studies have been conducted for the United States, starting with the work of Hogrefe et al. (2004), and these are reviewed by Jacob and Winner (2009). Results from ensembles of climate models are presented by Weaver et al. (2009). Consistently, models project increases for polluted U.S. regions ranging from 1 to 10 ppb in surface O_3 as a result of climate change over the coming decades, with largest increases found in urban environments and during peak pollution episodes. All models find consistently large increases in the northeast, but there is more disagreement in the southeast and west (Weaver et al., 2009). The climate-change impact on AQS violations will lessen if NO_x emissions decrease (Wu et al., 2008b), due in part to a decrease in the O_3 baseline (Murazaki and Hess, 2006). These high O_3 pollution events in the future climate are driven by local meteorology and local emissions, since the baseline O_3 changes are uncertain.

Impact of Non-Local Emissions Change on Local Air Quality Baseline O_3 will change as anthropogenic emissions of precursors around the world shift as a result of population and energy trends, industrialization of the developing world, and environmental controls in the developed world. NO_x emissions in China have doubled over the past decade (Zhang et al., 2008a), while emissions in the United States and Europe have decreased (van der A et al., 2008). If precursor emissions across North America and Europe continue to decrease in order to meet air quality goals, increasing emissions in the developing world could offset some of these benefits.

Model simulations based on the 2001 IPCC scenarios projected future increases in the surface O_3 baseline over the Northern Hemisphere of 2-7 ppb by 2030 (Prather et al., 2003), but more recent work corrects some of the IPCC assumptions and reduces precursor emissions (Dentener et al., 2005; Stevenson et al., 2006). Additional model studies using the uncorrected IPCC scenarios project a 2-6 ppb increase in baseline O_3 over the United States due to emissions outside North America by 2050 using one IPCC scenario (Wu et al., 2008a) and < 1 ppb increase with another lower-emissions IPCC scenario (Lin et al., 2008).

A more consistent prediction was that baseline O_3 will increase due to the increasing abundance of CH_4 (Prather, 2001). All of the 2001 IPCC scenarios predicted an increase in anthropogenic CH_4 emissions over at least the next several decades and an increase in baseline O_3 to accompany it, but this projection ran counter to the observations showing a leveling of CH_4 abundances since 1998 (Dlugokencky et al., 2003). Various explanations have arisen regarding the CH_4 trend (e.g., Bousquet et al., 2006; Chen and Prinn, 2006; Fiore et al., 2006), but there is no clear consensus as to the cause. In 2007, however, the annual CH_4 increases returned (Rigby et al., 2008). Given that most inventories project increasing anthropogenic CH_4 emissions (e.g., Wuebbles and Hayhoe, 2002), one must be prepared to accept parallel increases in baseline O_3.

> **Finding.** Projected climate change will lead to a warmer climate with shifts in atmospheric circulation. All of these changes have the potential to affect air quality, for better or worse:
>
> • higher water vapor abundances associated with a warmer climate drives down baseline O_3 abundances due to the more rapid photochemical destruction of O_3 in the marine boundary layer and the tropics;

- projections of more rapid stratospheric circulation increases the flux of O_3 into the troposphere, raising the baseline;
- warmer temperatures generate higher O_3 pollution levels from local emissions;
- for some regions, projected increase in wildfires can increase O_3;
- changing tropospheric circulation, such as the northward shift of storm tracks at mid-latitudes, may increase/decrease the number of stagnation events that lead to the worst pollution episodes; and
- increasing convective activity and lightning will enhance NO_x and the O_3 baseline.

In general these changes reduce the role of non-local emissions while enhancing that of local emissions in AQS violations.

Recommendation. Test and improve the simulation of air quality, including pollution episodes in both global and regional climate models. Ensure that the climate models include global atmospheric chemistry so that the model ensembles run for climate change assessments can also be used to evaluate changes in air quality. Analyze the climate statistics that are specific to air pollution: boundary layer height, stagnation episodes, intercontinental transport, stratosphere-troposphere exchange, lightning, wildfires, and other natural emissions of O_3 precursors.

Finding. In East Asia and much of the developing world O_3 precursor emissions are expected to increase rapidly over the next few decades and would likely raise the surface O_3 baseline in the United States by a few ppb. Methane increases as projected and recently observed, if continued over the next few decades, would additionally raise the O_3 baseline by a few ppb.

Recommendation. (1) Develop and improve the accuracy of emissions inventories for the current epoch as well as projections for the next decade. Include natural emissions and the development of interactive emissions models that depend on climate variables. (2) Establish a global international measurement strategy (e.g., through WMO/GAW) based on surface sites and airborne platforms that can identify changes in regional emissions of the ozone precursors: CH_4, NO_x, CO, VOCs, and relevant secondary compounds, such as PAN. (3) Merge these operational networks with satellite measurements and develop the modeling capability to provide reliable measures of emissions on a subcontinental scale. (See the discussion of integrated systems in Chapter 6.)

SUMMARY OF KEY FINDINGS AND RECOMMENDATIONS

Question: What is known about current long-range transport of ozone and its precursors?

Finding. Combining the evidence from observations and models and including our basic knowledge of atmospheric chemistry there is high confidence that human activities have raised the baseline levels of tropospheric O_3 in the Northern Hemisphere by 40-100 percent above preindustrial levels. Much of this increase can be directly attributed to anthropogenic emissions of ozone precursor species (CH_4, NO_x, CO, VOCs).

Finding. Baseline tropospheric O_3 abundances at many remote locations in the Northern Hemisphere have changed over the last few decades at rates ranging from hardly at all to as much as 1 ppb/yr. The cause of these changes is not clear.

Recommendation. The measurements documenting changes in baseline O_3 over the last few decades need to be systematically and collectively reviewed using consistent methods of analysis. Observations of O_3 and related trace species should be placed in perspective through global chemical-transport modeling that includes the history of key factors controlling tropospheric O_3: anthropogenic and natural emissions of precursors, stratospheric O_3, and climate. An analysis of the key uncertainties should be undertaken with an eye toward ensuring that future changes can be attributed with greater confidence.

Finding. Plumes containing high levels of O_3 and its precursors (NO_x, CO, VOCs) are observed downwind of the major industrial regions (and large wildfires) in North America, Europe, and Asia. They can be transported between continents and are observed in the free troposphere over affected regions but rarely at the surface due to dilution in the boundary layer.

Recommendation. Conduct focused research efforts that couple measurements with models to quantify the process of air exchange between boundary layer and free troposphere in order to better understand how free tropospheric O_3 enhanced by long-range transport is mixed to surface.

Finding. Baseline O_3 levels in the Northern Hemisphere are elevated by current anthropogenic emissions of NO_x, CO, and VOCs. Multimodel studies calculate that a 20 percent reduction in these emissions from any three of the four major industrial regions of the Northern Hemisphere will reduce surface O_3 in the fourth region by about 1 ppb on average, but with large spatial and seasonal variation. Results of similar magnitude are found for a 20 percent reduction in CH_4 emissions. Unfortunately, the range of results across this multimodel ensemble is comparable to the average result. These calculations were conducted for 2001 and in the following eight years these emissions have changed by 20 percent or more.

Recommendation. Since models provide the only practical method of attributing a fraction of surface O_3, and possible AQS violations, to different sources of pollution, a major effort must be made to calibrate, test, and improve these models with a wide range of atmospheric chemistry measurements ranging from individual events to climate statistics of air pollution. These tests must span the range of necessary modeling from global transport and stratosphere-troposphere exchange down to the urban airshed. While improving the models is a priority, realistically quantifying their uncertainty is key. Development of adjoint and ensemble modeling approaches will be important in assessing NAAQS compliance and intracontinental pollution under rules such as the U.S. EPA Clean Air Interstate Rule.

Question: What are the potential implications of long-range O_3 transport for U.S. environmental goals?

Finding. U.S. NAAQS violations (e.g., 8-hr average greater than 75 ppb) are caused primarily by a combination of regional emissions and unfavorable meteorology. These are augmented by a changing baseline and episodic events caused, for example, by wildfires, lightning NO_x, occasional stratospheric intrusions, and distant anthropogenic emissions. Most violations are only a few ppb above the standard; thus the increase in baseline O_3 since the preindustrial era driven by global pollution has contributed to these.

Finding. With high confidence we can state that elevated levels of O_3 occur over populated regions due to distant pollution and such increases are detrimental at some level to human health, agriculture, and ecosystems. A preliminary study finds that about 500 premature cardiopulmonary deaths could be avoided annually in North America by a combined 20 percent reduction in NO_x, CO, and VOC emissions from the other three major Northern Hemisphere industrial regions and, correspondingly, about 1800 in Europe. The uncertainty in these estimates is large, at least ± 50 percent, and reflects uncertainties in modeling both O_3 change and health effects.

Recommendation. Perform additional research to (1) attribute quantitatively the surface O_3 change to distant emissions; (2) specify the shape of the exposure-response curve (i.e., whether a threshold exists); (3) estimate avoided premature health burdens across a wider range of health outcomes; and (4) estimate avoided premature health burdens for particularly sensitive groups (i.e., children and others).

Question: What factors might influence the relative importance of U.S. and foreign contributions to O_3 change in the future?

Finding. Projected climate change will lead to a warmer climate with shifts in atmospheric circulation. All of these changes have the potential to impact air quality for better or worse:

• Higher water vapor abundances associated with a warmer climate drives baseline O_3 abundances down due to the more rapid photochemical destruction of O_3 in the marine boundary layer and the tropics.
• Projections of more rapid stratospheric circulation increases the flux of O_3 into the troposphere, raising the baseline.
• Warmer temperatures generate higher O_3 pollution levels from local emissions.
• For some regions, projected increase in wildfires can increase O_3.
• Changing tropospheric circulation, such as the northward shift of storm tracks at midlatitudes, may increase or decrease the number of stagnation events that lead to the worst pollution episodes.
• Increasing convective activity and lightning will enhance NO_x and baseline O_3.

In general these changes reduce the role of nonlocal emissions while enhancing that of local emissions in AQS violations.

Recommendation. Test and improve the simulation of air quality, including pollution episodes in both global and regional climate models. Ensure that the climate models include global atmospheric chemistry so that the model ensembles run for climate change assessments can also be used to evaluate changes in air quality. Analyze the climate statistics that are specific to air pollution: boundary layer height, stagnation episodes, intercontinental transport, stratosphere-troposphere exchange, lightning, wildfires, and other natural emissions of O_3 precursors.

Finding. In East Asia and much of the developing world O_3 precursor emissions are expected to increase rapidly over the next few decades and would likely raise the surface O_3 baseline in the United States by a few ppb. Methane increases, as projected and recently observed, if continued over the next few decades would additionally raise the O_3 baseline by a few ppb.

Recommendation. (1) Develop and improve the accuracy of emissions inventories for the current epoch as well as projections for the next decade. Include also natural emissions and the development of interactive emissions models that depend on climate variables. (2) Establish a global, international measurement strategy (e.g., through WMO/GAW) based on surface sites and airborne platforms that can identify changes in regional emissions of the ozone precursors: CH_4, NO_x, CO, VOCs, and relevant secondary compounds such as PAN. (3) Merge these operational networks with satellite measurements and develop the modeling capability to provide reliable measures of emissions on a subcontinental scale. (See the discussion of integrated systems in Chapter 6.)

3

Particulate Matter

THE COMPLEX NATURE OF PARTICULATE MATTER

Unlike most other pollutants, particulate matter (PM) cannot be characterized by the space- and time-variations of the mass concentrations of a single compound. Important factors influencing PM transport and its environmental and health effects include the following:

Size and Morphology   Environmental effects and lifetimes of particles vary with their size; thus, PM is classified by aerodynamic diameter. "PM_{10}" is the mass of particulate matter with diameters smaller than 10 µm, and "$PM_{2.5}$" or "fine PM" designates the fraction with aerodynamic diameters smaller than 2.5 µm; the "coarse" fraction is PM_{10} to $PM_{2.5}$. Most attention has been focused on the fine fraction because it affects health, visibility, and radiative forcing. With lifetimes on the order of days to weeks, fine particulate matter can undergo long-range transport, producing global and regional in addition to local impacts. While particles larger than 10 µm are also found in the atmosphere, rapid removal generally limits their lifetime to the order of hours, and as they are too large to be respirable, their health impacts are considered of minor importance. Particles are often assumed to be spherical, whereas this is only the case for nonaggregated, liquid particles. Although not generally characterized, even in intensive measurement campaigns, particle shape and phase can influence radiative properties and health impacts.

Chemical Composition   The varying chemical and physical nature of PM complicates the assessment of its impacts. Its composition depends on the emitting sources or particle precursors and also on atmospheric conditions. Some PM components—Including nitrate species, organic species, and water—are semivolatile and repartition between the gas and particle phase depending on environmental factors such as temperature, relative humidity, or the composition of the PM. For modeling and monitoring purposes the composition of dry atmospheric PM is generally reduced to a few major categories. Commonly identified components include sulfates, nitrates, organic carbon, elemental or black carbon, sea salt, soil or crustal material, and specific elements of interest, such as Pb. The health effects associated with exposure to these individual components are not well characterized and are likely to vary signficantly.

Primary and Secondary Sources   PM is emitted from both natural and anthropogenic sources, and its components are both primary (directly emitted) and secondary (formed in the atmosphere). Direct natural emissions come from wildfires, sea spray, and resuspension of organic matter such as leaf litter. The first of these produces primarily $PM_{2.5}$, while the latter two are mainly PM_{10}. Mineral dust has both natural and anthropogenic origins: it is lofted from arid and semiarid regions and can be mobilized by agricultural or construction activities. Its emission rates are especially susceptible to climate conditions. Combustion of fossil fuels and biofuels is a large primary anthropogenic source. Combustion processes are the only sources of black carbon, which together with "brown" carbon[1] has an important role in PM light absorption. Sources of secondary PM precursors (gases leading to particulate matter through atmospheric reactions) include gaseous vegetative emissions, motor vehicle emissions, and wood-smoke emissions. Reduced sulfur and nitrogen compounds are oxidized to the particulate components sulfate and nitrate, respectively. Ammonium is a common cation (positively charged ion) incorporated from the gas phase into the particle phase to neutralize these acid secondary species, although sodium, calcium, and other cations derived from sea salt or minerals are also often present.

Mass or Number?   Regulations and many measurement strategies have focused on characterizing the mass concentrations of PM, but for some health and welfare effects size or number concentrations may be the more relevant characteristic. This question is still unresolved. As discussed above PM differs from gaseous pollutants because it is a complex mixture with

---

[1] "Black carbon" refers to combustion-generated carbonacous aerosol that strongly absorbs visible light. "Brown carbon" refers to carbonaceous particles that are optically in between strongly absorbing black carbon and nonabsorbing organic carbon, formed largely by inefficient combustion of hydrocarbons.

wide variations in chemical composition and particle sizes. Their chemical composition reflects the particles' origin as well as processing that has occurred during their lifetimes in the atmosphere. This report focuses mainly on particles with aerodynamic diameters below 2.5 micrometers ($PM_{2.5}$) because they can be transported over much longer distances than larger particles, and because they have stronger effects on the health and environmental impacts of interest. Further, we focus on the regulated and therefore most available observed properties, mass concentrations, and bulk chemical compositions of PM, noting that other properties may assume more importance in the future as understanding of PM effects increases and regulations are accordingly revised. For example, PM acts as a vehicle for the atmospheric transport and inhalation of toxic substances such as heavy metals (Murphy et al., 2007) and PAHs (see Chapter 5).

The chemical composition of $PM_{2.5}$ offers clues to its sources and atmospheric chemistry, and chemical characterization is critical to understanding the environmental impacts of present and future sources. Numerous measurements of PM composition have been published (Kim et al., 2000; Edgerton et al., 2006; Ondov et al., 2006). These are complemented by measurements of various components of PM emitted by specific sources (Andreae and Merlet, 2001; Watson et al., 2001; EPA, 2006b), including advanced measurements that characterize individual particles (Sodeman et al., 2005).

Mass concentrations of each species and size fraction are measured and reported using operational definitions and techniques (generally filter based) (e.g., Pitchford et al., 2007), which give an idea of the overall dry composition of the sampled size fraction. It is important to bear in mind that water can be the dominant (and unmeasured) PM constituent above ~ 90 percent relative humidity and sampling and analysis difficulties are known to be present for elemental carbon, organic carbon, and nitrate species, making their routine quantification less certain. More sophisticated measurements that are applied in intensive field studies are able to provide better characterization of the PM than that possible from filters, such as the mixing state of species within individual particles (Murphy et al., 2006) and the presence of trace quantities of toxic species (Moffet et al., 2008).

In this chapter we favor the use of the word "baseline" to indicate a potentially observable quantity that represents the cleanest 20 percent of daily surface observations. The word "background" implies an atmospheric concentration that occurs in the absence of modern anthropogenic sources; while models may be used to provide such a distinction, measurements are presently indifferent to this fractionation.

SOURCES OF PARTICULATE MATTER

Figure 3.1 shows the major sources of primary emissions of $PM_{2.5}$ in the United States. Major components of these particles include sulfate, nitrate, black carbon, organic carbon, and mineral dust; the latter two are themselves complex mixtures. Figure 3.2 summarizes the global sources of major non dust PM components, including sources leading to secondary production of sulfate, nitrogen, and organic PM components. Secondary organic aerosol (SOA) precursor gases may originate naturally from plants (biogenic emissions) (Guenther et al., 2000), from open burning of biomass (Grieshop et al., 2008), or from energy-related sources (Robinson et al., 2007). Neither the reactions nor the exact compounds leading to formation of SOA are fully understood (Donahue et al., 2009), as demonstrated in recent comparisons of modeling and observations (e.g., Heald et al., 2006).

We note that the commonly measured major components included in Figure 3.2 do not fully characterize the often complex chemical nature of PM, and some of the minor species present are known to be toxic to humans or the environment. For example, the organic carbon fraction may contain POPs (see Chapter 5), a category that encompasses pesticides and polycyclic aromatic hydrocarbons, among other toxic species. Many metals have been detected in PM, including lead, zinc, mercury, and vanadium (e.g., Moffet et al., 2008). While these elements generally contribute only minimally to total

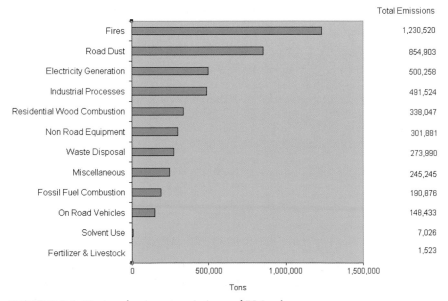

FIGURE 3.1 National primary emissions of $PM_{2.5}$ by sector.
SOURCE: EPA, http://epa.gov/air/emissions/pm.htm.

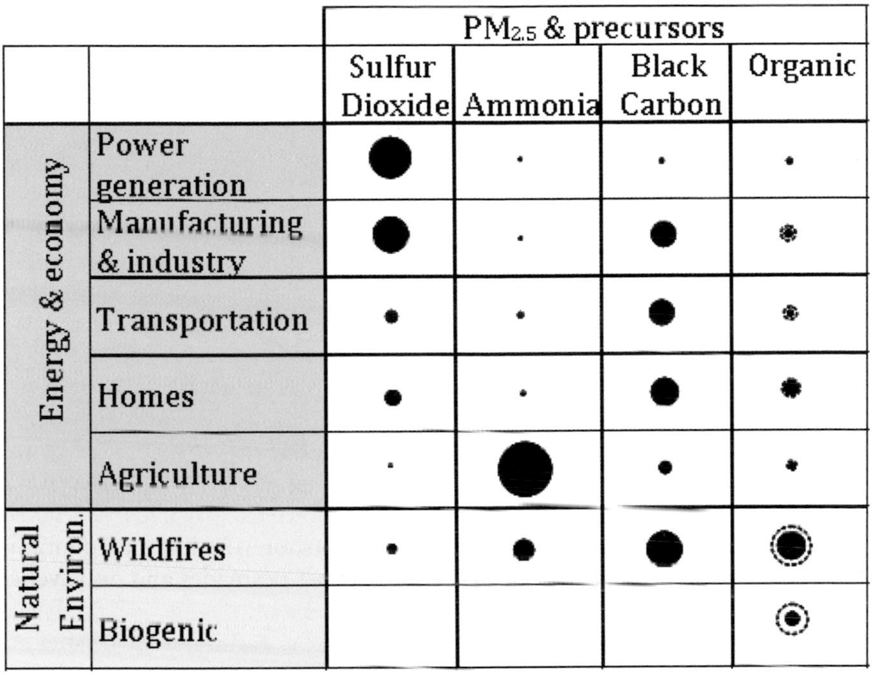

		PM$_{2.5}$ & precursors			
		Sulfur Dioxide	Ammonia	Black Carbon	Organic
Energy & economy	Power generation	●	·	·	·
	Manufacturing & industry	●	·	●	●
	Transportation	·	·	●	●
	Homes	·	·	●	●
	Agriculture	·	●	·	·
Natural Environ.	Wildfires	·	·	●	◉
	Biogenic				◉

FIGURE 3.2 Global sources of PM$_{2.5}$ components and precursors. Areas of black dots are proportional to fraction of each compound originating from individual sources; that is, dots in each column sum to the same total area. Sulfur dioxide (a precursor) leads to sulfate, while PM nitrogen species result from complex reactions involving ammonia and other nitrogen compounds, including organics. In the column for organic carbon, dashed circles show the approximate contribution of secondary organic aerosol formed from precursors. Values are from Bond et al. (2004), Cofala et al. (2007), and Heald and Spracklen (2009); secondary organic aerosol values are adapted from de Gouw and Jimenez (2009).

PM mass, the trace amounts carried by particulate matter can represent biologically significant levels and their presence in fine particulate matter means they may be able to lodge deeply in the lungs if inhaled.

PARTICULATE MATTER IMPACTS

Human Health Impacts   Exposure to elevated levels of PM has long been associated with adverse human health impacts, and is a major source of morbidity and mortality worldwide. A European study using exposure-response functions for a 10 µg m^{-3} increase in PM$_{10}$ found that this increase accounted for 6 percent of total mortality, as well as more than 25,000 new cases of chronic bronchitis in adults, more than 290,000 episodes of bronchitis in children, more than 500,000 asthma attacks, and more than 16 million person-days of restricted activities in Austria, France, and Switzerland (Künzli et al., 2000). Health effects have been observed at all exposure concentrations, indicating a wide range of susceptibility and that some people are at risk even at the lower end of observed concentrations. There is convincing and consistent evidence that short- and long-term exposures to particulate matter can cause a wide range of adverse health effects. Effects related to short-term exposure include lung inflammatory reactions, respiratory symptoms, adverse effects on the cardiovascular system, increase in medication usage, and increase in mortality. Effects related to long-term exposure include increase in lower respiratory symptoms, reduction in lung function in children and adults, increase in chronic obstructive pulmonary disease, and reduction in life expectancy due mainly to cardiopulmonary mortality but also to lung cancer (WHO, 2006).

Figure 3.3 shows summary estimates for relative risks for mortality caused by different air pollutants. Particulate mass, particularly the mass in the smaller particles (e.g., $PM_{2.5}$), show the highest relative risk for mortality. It remains unclear what components or characteristics are responsible for observed adverse health effects, with hypotheses including the presence of metals, sulfates, and other acidic species, or simply size (smaller than 0.1 μm) regardless of composition. Studies are inconsistent in attributing the health effects, whether to gaseous co-pollutants, particles themselves, trace elements, or toxic substances such as benzene and manganese (Samet and Krewski, 2007).

Recent studies estimated a 0.21 percent increase in deaths per day per 10 μg m^{-3} increase in PM_{10} exposure, and an increase of approximately 4-8 percent in long-term risk of death for each 10 μg m^{-3} rise in annual $PM_{2.5}$ concentrations (Kaiser, 2005; Krewski et al., 2009). Comparing data on life expectancy, socioeconomic status, and demographic characteristics for 51 U.S. metropolitan areas with data on $PM_{2.5}$ concentrations for the late 1970s to early 1980s with the late 1990s to early 2000s, Pope III et al. (2009) concluded that a decrease of 10 μg m^{-3} in the concentration of $PM_{2.5}$ was associated with an estimated increase in mean life expectancy of 0.61 ± 0.20 year. A similar analysis in Europe, using a concentration-response function of an increase in risk of all-cause mortality by 6 percent per 10 μg m^{-3} of $PM_{2.5}$, found that current exposure to PM from anthropogenic sources leads to an average loss of 8.6 months of life expectancy (WHO, 2006). The total number of premature deaths attributed to PM exposure was about 348,000 in 25 countries. Effects other than mortality, including some 100,000 annual hospital admissions, were also attributed to exposure. The study further concluded that PM from long-range transport

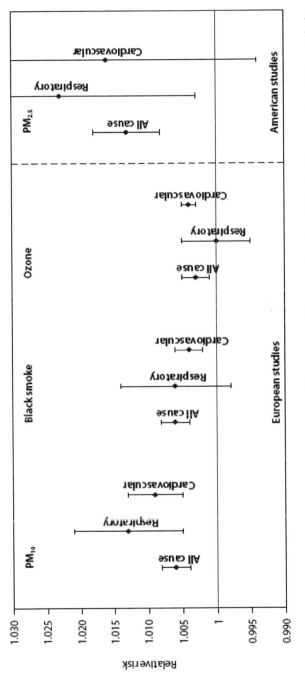

FIGURE 3.3 Summary of relative risks for mortality by different air pollutants, related to a 10 μg/m3 increase in pollution including 95 percent confidence intervals. PM_{10} includes $PM_{2.5}$. A relative risk of 1.005 indicates a 0.5 percent increased risk of mortality. SOURCE: WHO (2004).

of pollutants contributes significantly to these effects. Recently Corbett et al. (2007) estimated that shipping-related PM emissions contribute ~ 60,000 excess deaths annually worldwide, primarily in coastal regions, which face greater exposure to emissions from port areas, and generally have higher population density (see Chapter 6).

Ecosystem Impacts   Wet and dry deposition of particulate-phase sulfate and nitrate species constitute major contributions to the atmospheric inputs of sulfur, nitrogen, and acidity to ecosystems. Particles can also serve as vehicles for the atmospheric transport and ultimately deposition of other pollutants, such as heavy metals and POPs. The interactions of atmospheric aerosols with solar radiation can affect rates of photosynthesis in complex ways (Cohan et al., 2002).

Visibility and Radiative Forcing Impacts   Submicron particulate matter is especially efficient at extinction of visible and ultraviolet radiation and represent the major contribution to visibility reduction in both rural and urban settings. This same mechanism is the basis for PM direct radiative forcing effects, that is, their impact on global radiation budget. The impacts of PM on radiative forcing, and the response of that forcing to future changes in PM emissions and lifetimes, are fundamentally different from those of greenhouse gases (GHG). First, while most GHGs are long-lived, PM has a short lifetime, so its influence on radiative budgets is more variable in space and time. Second, the direct radiative impacts of PM can be either warming or cooling, depending on the optical properties.

Global average impacts   The Intergovernmental Panel on Climate Change (Forster et al., 2007a) produces regular assessments of radiative forcing (RF) and climatic impact. Aerosols generally have a net negative climate forcing because they reflect sunlight away from Earth. Exceptions with positive RF are strongly absorbing aerosols such as black carbon, or aerosols with even a small amount of absorption that are located above clouds. Because the sign of forcing (positive or negative) varies with different types of aerosol particles, mitigation (or increased emissions) of these different aerosols will have very different impacts on climate.

IPCC (2007) most recently estimated the direct RF of all anthropogenic aerosols as $- 0.5$ W/m^2. Compared with greenhouse gas forcings of $+ 2.6$ W/m^2, this value is smaller but not insignificant. If recent findings regarding the ubiquity of organic aerosol (Heald et al., 2005) were linked with rapid secondary organic aerosol formation from anthropogenic sources (e.g. Robinson et al., 2007), the RF estimates would become slightly more negative. Several indirect links between PMs and climate have been postulated, because particles serve as sites for the formation of liquid and ice-cloud particles. These include PM-induced modifications to cloud frequency, altitude, lifetime, microphysical and radiative properties, and effects on precipitation. Pollution aerosols are estimated to modify clouds to have smaller, more reflective droplets, with an additional negative forcing estimated at about -0.7 W/m^2 (IPCC, 2007) with a ~ 50 percent uncertainty.

Regional impacts Global average forcing is only a partial measure of aerosol-climate interactions. Aerosol RF near source regions can be greater than global average RF by an order of magnitude, and RF reflects changes at the top of the atmosphere, ignoring redistribution of energy within the atmosphere (Satheesh and Ramanathan, 2000). Regional impacts of aerosols may include shifts in rainfall distribution (Rotstayn and Lohmann, 2002; Wang, 2007) or energy available to drive the hydrologic cycle (Meehl et al., 2008). Thus, the net aerosol impact on climate is not neatly represented by radiative forcing (NRC, 2005).

Climatic impacts of transported PM The source-receptor approach described by HTAP-TF (2007) might also be used to indicate the impact of emissions in one region on RF, optical depth, or column burden in another region. Aerosol optical depths above downwind continents can be ~ 10-20 percent of those in source regions (Koch et al., 2007). Model calculations of Reddy and Boucher (2007) predicted that 20-30 percent of the black carbon column loading in East Asia, North America, and Europe was imported from other regions. RF by aerosol and gaseous emissions from isolated regions and sectors can be of the order of 100-400 mW/m^2 (Shindell et al., 2008a). These perturbations are small but not insignificant; radiative impacts of one region on another region are of the same magnitude as global average aerosol forcing.

Despite the fact that aerosol PM concentrations vary regionally, recent studies on the impact of aerosol distributions on climate suggest that this response should be considered in a global context. Levy et al. (2008) and Kirkevåg et al. (2008) have found that climate simulations using aerosol loadings projected to the year 2100 using the SRES A2 scenario produce regional patterns of surface temperature warming that do not follow the regional patterns of changes in aerosol emissions, tropospheric loadings, or radiative forcing. Rather, the regional patterns of warming from aerosols are more similar to the patterns for well-mixed greenhouse gases.

Sensitive regions For regions with no internal sources of pollution, intercontinental transport from industrialized regions represents the only anthropogenic impact. For example, transported aerosols and other pollutants contribute to haze in the Arctic (Law and Stohl, 2007), which has both warming and cooling components (Quinn et al., 2008). Absorbing aerosols in the haze when deposited on snow may hasten the onset of spring melt (Flanner et al., 2007). The connection between transported pollution and impacts on sensitive regions is an area of active research. Transport from lower latitudes to the Arctic is particularly uncertain in comparison with other source-receptor relationships (Shindell et al., 2008b).

THE REGULATORY CONTEXT FOR CONTROL OF PM

The import of PM into the United States has the potential to affect the ability to achieve compliance with U.S. regulations, which are aimed at both limiting ambient levels of PM to protect human health and welfare and at controlling emissions from new and existing sources. At the present time welfare-based standards for pollutants apply only to visibility impairment and potential crop damage and not to direct or indirect radiative forcing impacts.

Surface Concentrations: Primary Standards　Current U.S. health-based (primary) NAAQS for ambient concentrations of particulate matter focus on dry, unspeciated PM mass concentrations and exist for both PM_{10} and $PM_{2.5}$. The most recent revision of the U.S. particulate matter standard, promulgated in 2006, tightened the 24-hr fine particle ($PM_{2.5}$) standard from 65 µg m^{-3} to 35 µg m^{-3} and retained the annual fine particle standard at 15 µg m^{-3}. To be in compliance with the annual standard the 3-yr average of the weighted annual mean $PM_{2.5}$ concentrations from single or multiple community-oriented monitors must not exceed 15 µg m^{-3}, while to attain the 24-hr standard, the 3-yr average of the 98th percentile of 24-hr concentrations at each population-oriented monitor within an area must not exceed 35 µg m^{-3}. The existing 24-hr PM_{10} standard of 150 µg m^{-3} was retained, but the annual PM_{10} standard was revoked due to the lack of clear evidence linking long-term exposure to PM_{10} with adverse health impacts. Figure 3.4 compares the 24-hr $PM_{2.5}$ ambient standards for the United States, Canada, Mexico, and the World Health Organization. U.S. standards are higher than WHO-recommended levels, and may undergo further downward revision as epidemiological data continue to be collected and analyzed; the relative role of non-U.S. sources will assume more importance as standards are tightened.

We note here that even relatively small fractional contributions of PM, as may occur from long-range transport of PM, can be significant in national control strategies. For example, the U.S. Prevention of Significant Deterioration (PSD) Act and Clean Air Interstate Rule (CAIR) specify allowable incremental emissions for new U.S. sources. Figure 3.4 shows the proposed allowable 24-hr $PM_{2.5}$ emissions increment (2 µg m^{-3}) for

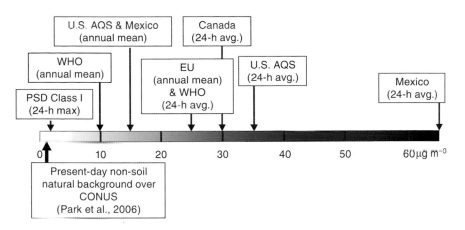

FIGURE 3.4 Comparison of current 24-hr health-based PM$_{2.5}$ standards for the indicated countries, and U.S. allowable 24-hr emissions increment for Class I areas under the Prevention of Significant Deterioration rule.

new sources impacting Class I areas under PSD; the annual mean allowable increment is 1 µg m^{-3}. Under CAIR, EPA considers an upwind state to contribute significantly to a downwind nonattainment area if the state's maximum contribution to PM$_{2.5}$ in the area is ≥ 0.2 µg m^3.

Surface Concentrations: Secondary Standards U.S. secondary standards are set to protect human welfare, including visibility. In contrast to the NAAQS approach, which requires all areas of the nation to demonstrate that ambient pollutant concentrations are below a specified standard, the regional haze rule (passed in 1999) addresses only national parks and wilderness areas. State and federal agencies are required to improve visibility in these areas toward "natural background conditions" by 2064. Background conditions are those unaffected by anthropogenic emissions, so these may include impacts from sources such as fires and dust, including PM that has undergone long-range transport. Rising anthropogenic PM emissions from other countries that impact U.S. PM concentrations could present significant roadblocks to the achievement of designated background visibility conditions (Park et al., 2006), a complication not clearly addressed in the existing U.S. regulatory framework.

Emissions Standards Primary PM emissions from on-road and non-road mobile sources have been specifically targeted for reductions in the United States. The EPA established progressively more stringent mobile-source PM emission standards beginning in the mid-1970s for on-road vehicles and in the early 1990s for non-road engines and equipment. With the designation of diesel exhaust as a likely human carcinogen, a particular focus in recent years has been establishing emission standards for diesel engines. Diesel PM is found primarily in the $PM_{2.5}$ fraction, and studies show that as sulfur is removed from diesel fuel, the sizes of emitted particles become smaller. Black carbon is a main components of diesel exhaust and its ability to absorb shortwave radiation contributes a positive radiative forcing.

Other emission-based U.S. regulations include acidic deposition regulations, which are tied to PM concentration levels because the regulated pollutants—sulfur dioxide and nitrogen oxides—are precursors of secondary PM. Reductions in these species achieved as a result of acid deposition regulations have already resulted in improvements in ambient PM concentrations in many parts of the United States (see Chapter 1). International shipping activity is an example of currently unregulated emissions that have the potential to significantly affect PM in some regions. For example, auxiliary diesel engines operating within 24 nautical miles of the ports of Long Beach and Los Angeles are believed to be an important source of PM, nitrogen oxides, and sulfur to the southern California region (Minguillón et al., 2008), accounting for 10-44 percent of the non-sea-salt sulfate found in fine PM in coastal California (Dominguez et al., 2008). Corbett et al. (2007) estimated annual average contributions to $PM_{2.5}$ from international shipping operations as high as 2 µg m^{-3} in some heavily traveled coastal regions (see Chapter 6 for further discussion of the role of shipping emissions in U.S. air quality).

Current Non-U.S. PM Policies   Future trends in PM emissions in various regions around the globe are tied to the air quality standards in those regions. This linkage occurs because not only the levels of emissions but also the choices of pollutants and sources to target for controls in national emission plans are driven by air quality standards. The PM standard is a good example and as shown in Figure 3.4 the $PM_{2.5}$ standard varies by country. Furthermore, the United States is the only country in the world with a visibility standard, and many countries, including most in Asia, do not have a $PM_{2.5}$ standard. This is important, as smaller particles (such as $PM_{2.5}$) can be transported over longer distances than larger particles, yet PM_{10} standards may emphasize sources that emit primarily coarse particles.

KEY PM TRANSPORT PATTERNS AND EFFECTS ON SURFACE CONCENTRATIONS

Local and regional emissions are responsible for most PM concentrations that exceed air quality standards, but it is now recognized that emissions and transport at the intercontinental and global scales can lead to episodic spikes in concentrations that cannot be attributed solely to U.S. emissions. Seasonal transport patterns are discussed in Chapter 1, and Appendix B discusses the various mechanisms that influence the vertical mixing of pollutants. Although there are many parallels between the patterns of transport of ozone and PM into and out of the United States, one important difference is that emissions of PM are efficiently depleted by deposition processes, especially wet deposition (e.g., Zhao et al., 2008 and numerous earlier references), resulting in low and highly variable baseline concentrations. Depositional losses are lowest along transport pathways occurring at higher altitudes and having faster windspeeds, and thus such air masses are responsible for transporting PM mass to the United States in concentrations that clearly exceed expected baseline conditions. This transport has been unambiguously identified primarily in large-scale dust plumes but also occasionally in plumes from wildfires or industrial pollutants. Although relatively infrequent, such extreme events can nonetheless be highly policy relevant, since current U.S. standards allow only a limited number of exceedances to be omitted from averages used to determine compliance with the NAAQS. These events also highlight the possibility that more persistent but less readily discerned transport may be occurring at other times, contributing small but potentially important incremental increases to surface PM concentrations.

Based on analyses of transport patterns and of the nature and composition of PM, extreme events clearly attributable to non-U.S. sources have been documented. Here we summarize present knowledge of transport pathways and patterns of PM into the United States, as discussed extensively in HTAP-TF (2007).

Asian Dust and Pollution Plumes containing PM originating from Asian sources primarily affect the western United States. Transport that is rapid and dry enough to retain PM in transit across the Pacific Ocean occurs in the midtroposphere (see Figure 1.2). In the continental United States the enhanced PM from these plumes is most clearly observed at higher-elevation sites (e.g., VanCuren and Cahill, 2002). Episodic high-concentration dust events are most frequently observed at the surface in springtime because of increased Asian dust mobilization and favorable transport pathways during that season (Wells et al., 2007), although some studies suggest that Asian emissions are transported annually to elevated sites across the United States year-round (VanCuren and Cahill, 2002; VanCuren, 2003; VanCuren et al., 2005). A few extreme events of Asian dust transport have been documented, including the large April 1998 plume (Husar et al., 2001) that was tracked across the United States and an even larger plume that was observed in April 2001 and was responsible for significant increments in PM_{10} (Jaffe et al., 2003c). Even a dust cloud from China, lofted to the upper troposphere, was transported more than one full circuit around the globe (Uno et al., 2009).

A recent analysis of surface PM data in the western United States, combined with satellite data, has identified significant interannual variations in the amount of transported Asian dust and pollution (Fischer et al., 2009). This analysis indicates that variations in transport and dust emissions could explain ~ 50 percent of the interannual variations in $PM_{2.5}$ concentrations at IMPROVE sites in the western United States. In high dust years average $PM_{2.5}$ concentrations at these sites were 3.7 µg m^{-3} (spring average), compared with 2.2 µg m^{-3} in low-dust years (Fischer et al., 2009). Some instances of elevated nondust PM have also been reported at West Coast sites (e.g., Jaffe et al., 2005b) and attributed to sulfur and other Asian industrial emissions (Bertschi and Jaffe, 2005; Weiss-Penzias et al., 2006, 2007; Hadley et al., 2007). Results from spring 2006 showed that episodic long-range transport events from Asia can contribute more than 1.5 µg m^{-3} of aerosol sulfate to coastal western Canada surface concentrations. The springtime mean contribution of the long-range transport of sulfate from Asia to surface PM levels in western Canada is estimated to be ~ 0.3 µg m^{-3} (van Donkelaar et al., 2008; Heald et al., 2006).

North African Dust   Dust from North African sources is mobilized year-round (Prospero et al., 2002), but transport pathways vary seasonally. The direction of westward North African dust transport is tied to the shifts in position of the Bermuda high, with plumes affecting Amazonia in the Northern Hemisphere winter, and transport shifting northward during the summer months. Summertime dust events, with transported dust concentrations generally greater than 10 µg m^{-3}, have been documented in the southeastern United States, occasionally extending into Texas and the mid-Atlantic states (e.g., Perry et al., 1997). NAAQS exceedances that may have occurred because of the incremental input of North African dust have been documented in the southeastern United States (Prospero, 1999a,b; Prospero et al., 2001). Liu et al (2009) uses source-receptor relationships to quantify the health impacts of the intercontinental transport of dust.

Asian and Canadian Wildfires   The role of CO emissions from Siberian wildfires in influencing interannual variations in the Northern Hemispheric CO budget was demonstrated by Wotawa et al. (2001). This work highlighted the large potential perturbations to tropospheric chemistry, including PM concentrations, that are driven by high-latitude wildfires (Wotawa and Trainer, 2000; Morris et al., 2006). PM from Canadian wildfires has been observed in the Great Lakes region and linked to air quality degradation (Al-Saadi et al., 2005). A recent extreme event of smoke transport from Canadian fires into Washington, DC, has been documented (Colarco et al., 2004). $PM_{2.5}$ NAAQS were close to being exceeded as a result of this transport (Bein et al., 2008). Emissions from the large 2003 Siberian fires were transported to the U.S. Pacific Northwest and contributed significantly to surface PM concentrations and to an 8-hr O_3 average greater than 90 ppbv (Jaffe et al., 2004). These fires also resulted in significant aerosol loading at the surface, with submicron aerosol scattering coefficients of up to 80 Mm^{-1}, or approximately 27 $\mu g\ m^3$ (Bertschi and Jaffe, 2005).

Mexican and Central American Agricultural and Wild Fires   Agricultural burning occurs each spring (April and May) in Mexico and Central America (Mendoza et al., 2005). Annual but variable impacts on PM concentrations in national parks across the southwestern United States have been documented (Gebhart et al., 2001; Park et al., 2003), but the impacts of these sources on NAAQS exceedances in the region are unknown (Choi and Fernando, 2007). In some years drought and weather conditions have caused burns to erupt into uncontrolled wildfires, and these extreme events are more readily identified in observations. For example, in 1998 shifting weather patterns clearly transported smoke from such wildfires into the United States (Kreidenweis et al., 2001; Park et al., 2003; In et al., 2007), where it was tracked northward and into the eastern United States (Peppler et al., 2000; Falke et al., 2001).

Transport and Deposition into U.S. Arctic Regions.   Short-lived species such as PM have been implicated in the magnification of Arctic warming (Quinn et al., 2008) through their direct and indirect effects on the radiation budget and through deposition of absorbing PM (black carbon) to snow and ice surfaces, where they can accelerate melting (Shindell, 2007). Transport and deposition of PM to the Arctic depends on the season, with varying contributions from the transport of European, East Asian, South Asian, and North American pollution, as well as dust and wildfire emissions (Quinn et al., 2008). The results from a coordinated comparison of 17 models (Shindell et al., 2008b) suggested that European emissions dominate aerosol transport to the surface in the Arctic, whereas East Asian emissions dominate in the upper troposphere.

MODELING AND ATTRIBUTION OF PM TRANSPORT AND TRENDS

Trends in PM Concentrations   Data from monitoring networks have been analyzed to attempt to discern trends in PM surface concentrations. Prospero et al. (2003) found that trends in non-sea-salt sulfate concentrations from a 20-year PM record from Midway Island in the North Pacific followed trends in SO_2 emissions from China, suggesting a similar trend might be expected in PM imported from Asia to the United States. In contrast, the analyses of Jaffe et al. (2005b) using PM data from four sites in the western United States showed no apparent long-term trend over a similar time period. In general, trend analyses of IMPROVE data (e.g., Malm et al., 2002) are either inconclusive or show trends that appear to be more closely tied with changing U.S. emissions (Schichtel et al., 2001). For example, DeBell et al. (2006) found 10-year (1989-1999) statistically significant decreasing trends in sulfate concentrations in Class I areas throughout much of the contiguous United States, linked to decreasing U.S. SO_2 emissions. A recent study by Wang et al. (2009b) analyzed visibility datasets from around the globe and found evidence for a global increase in aerosol optical depth from 1973 to 2007, despite a net decrease over Europe over this same period. The conclusions of DeBell et al. (2006) for surface PM are also different from this global estimate: they reported statistically-significant improving trends from 1995 to 2004 for the best visibility (generally lowest-PM) days at 17 IMPROVE sites in the western United States, including Alaska, suggesting baseline PM concentrations in those locations were not increasing over that time period.

Space-Based Observations   Satellite imagery has long been used to observe transport of PM such as pollution, smoke, and dust plumes from space, but only recently have attempts been made to use remote sensing products to obtain quantitative estimates of transported PM mass. Yu et al. (2008) combined MODIS aerosol retrievals with analyzed wind fields to estimate fluxes of pollution (nondust) PM leaving Asia and arriving in North America. They further showed that these estimates were very similar to those computed independently from a global model, and the imported fluxes are about 15 percent of local emissions from the United States and Canada, as also estimated by Chin et al. (2007). Yu et al. (2008) note large uncertainties on the order of a factor of 2 and stress that the estimates are column amounts and cannot be immediately applied to estimate effects on surface air quality. One complication is that water can represent the dominant mass fraction of $PM_{2.5}$, and its concentration in PM varies strongly with aerosol composition and ambient relative humidity, making it difficult to account for its presence from space-based observations alone. There is also increasing interest in using satellite observations of aerosol optical depth to infer information regarding surface $PM_{2.5}$ concentrations. Current experience suggests that $PM_{2.5}$ surface concentrations are proportional to satellite aerosol observations for regions where aerosol composition and vertical profiles are relatively stable (e.g., eastern United States), but this relationship weakens where aerosol composition and vertical profile have large variations (e.g., western United States) (Al-Saadi et al., 2005). Martin (2008) discusses application of remote sensing to monitor surface air quality, including a review of current techniques and criteria for future instruments aimed at this problem.

Modeling of PM Concentrations As indicated above, it remains difficult to evaluate the influence of non-U.S. sources on PM solely from present-day observations, particularly to discern those sources that might contribute relatively small increments to surface PM concentrations. Extreme events are the most readily identified and documented from observational datasets and exhibit large interannual variability due to large-scale influences, such as drought and general circulation changes (e.g., Prospero and Lamb, 2003; Fischer et al., 2009). Nevertheless, because monitoring sites are fixed and transport altitude and latitude can vary, observations are likely to miss even extreme events if monitoring sites are not located directly in the transport pathway. Modeling can help to fill in observational gaps and attribute PM increments to source regions. CTMs provide a means to estimate four-dimensional aerosol distributions based on an emission distribution and the governing meteorological fields. There are several recent activities to compare and evaluate PM predictions using CTMs, including the AEROCOM study designed to evaluate the main uncertainties in global aerosol models (Kinne et al., 2006; Schulz et al., 2006; Textor et al., 2006), the Model Intercomparison Study in Asia (MICS-Asia) (Carmichael et al., 2008a), and the HTAP study (HTAP-TF, 2007). There are also reviews focused on specific components of PM, for example, the review of secondary organic aerosol modeling by Kanakidou et al. (2005), and the intercomparison of dust predictions (Uno et al., 2006).

The treatment of PM in CTMs has become more complex in recent years (Kinne et al., 2006). Many models now describe PM as composed of several chemical species that are distributed among particles of diameters from nanometers to several microns. Most of the model comparisons have focused on model-to-model comparisons with limited in model-to-observation comparisons. This is due in large part to the limited number and sparse global coverage of PM composition measurements. The most extensive comparisons with measurement have focused on global datasets of aerosol optical depth from satellites and Sun photometers. Interestingly, the total aerosol optical depth in most models is comparable to observation-based estimates (Kinne et al., 2006), whereas the underlying aerosol fields of sulphate, organic matter, dust, and sea salt showed a much larger variation. Thus speciated PM observations are needed to better evaluate and constrain models.

In general, the predictive skill of CTMs for PM is poorer than that for ozone. For example, the skill of seven regional CTMs to predict surface $PM_{2.5}$ and ozone was recently evaluated against observations collected during August and September 2006, through the AIRNow network

(Aerometric Information Retrieval Now) in eastern Texas and adjoining states (McKeen et al., 2009). Ensemble O_3 and $PM_{2.5}$ forecasts created by combining the seven separate forecasts with equal weighting and simple bias-corrected forecasts were also evaluated in terms of standard statistical measures, threshold statistics, and variance analysis. For O_3 the models and ensemble generally show predictive skill relative to persistence for the entire region but fail to predict the highest O_3 events in the Houston region. For $PM_{2.5}$ none of the models nor their ensemble showed skill relative to persistence, and the models showed significant bias.

There are many reasons for the difficulty in prediction of PM and the large diversity in the individual PM component fields predicted by different models. For example, there are large uncertainties in the emissions of particles and their precursors (as discussed already in this chapter). In addition, there remain gaps in our understanding of the chemistry of PM formation and the rates by which PM is removed by dry and wet processes, stemming in large part from the difficulty in modeling cloudiness and precipitation. While the relative sources of uncertainty vary from species to species, in general the uncertainties (normalized: uncertainty/mean) are large by factors of 2 to 6 (Bates et al., 2006).

Only a few studies have explored source-receptor relationships for aerosol emissions and burdens on continental scales. The HTAP report (HTAP-TF, 2007) summarizes this work and presents the most comprehensive multimodel analysis of source-receptor relationships on hemispheric scales. Figure 3.5 summarizes the results from the model calculations in terms of the extent to which 20 percent perturbations in local and distant anthropogenic emissions of aerosol precursors and aerosol primary emissions contribute to mean surface $PM_{2.5}$, total sulfate deposition, and sulfate column loadings over several receptor source regions: North America (NA), Europe (EU), South Asia (SA), and East Asia (EA). It should be pointed out that these results are the annual and NA mean values. The results for a specific location and period of time may be significantly different from the mean. In Figure 3.6 these same data are presented as import sensitivities, defined as the sum of the changes in the quantity due to 20 percent perturbations in anthropogenic emissions in all other regions, divided by the change in that quantity due to 20 percent changes in domestic emissions. The +/– 1 standard deviation between the model results (10 models) are also shown and indicate that there is a large diversity among models in the attribution of the role of distant sources.

For North America the relative contribution from all distant sources (Europe, East Asia, and South Asia added together, divided by the contribution from NA sources) to surface $PM_{2.5}$, total sulfate deposition, and sulfate column loadings is estimated as 5, 9, and 33 percent, respectively. Shown in Figure 3.7 is the relative contribution of NA emissions to surface $PM_{2.5}$

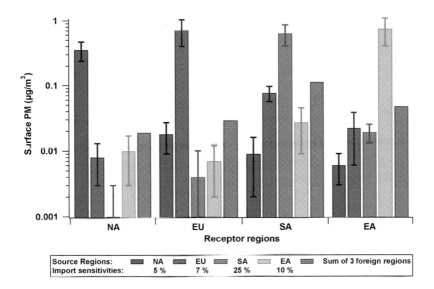

FIGURE 3.5 Increments in modeled surface PM$_{2.5}$ concentrations in several regions (NA = North America, EU = Europe, SA = South Asia, EA = East Asia), computed for 20 percent increases in emissions from the indicated source regions (colored bars). The error bars indicate +/- 1 one standard deviation among the 10 participating models. Note the log scale on the y-axis.

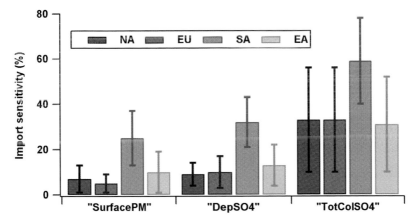

FIGURE 3.6 Import sensitivities for the various regions in Figure 3.5, defined as the sum of the changes in the quantity due to 20 percent perturbations in anthropogenic emissions in all other regions, divided by the change in that quantity due to 20 percent changes in domestic emissions, expressed as a percentage. "SurfacePM" = surface concentrations of PM2.5; "DepSO4" = sulfate deposited to the surface; "TotColSO4" = total column sulfate loading.

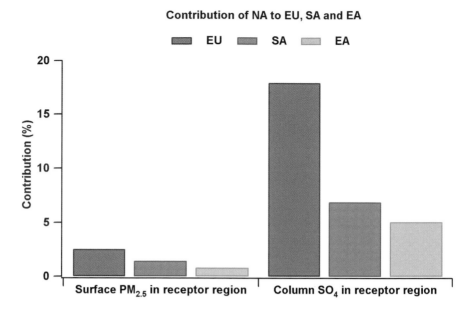

FIGURE 3.7 The contribution to surface PM$_{2.5}$ and sulfate column amounts (expressed as a percentage) of North American emissions relative to domestic emissions in the receptor region (e.g., for EU = ΔNA/ΔEU).

and sulfate column amounts due to changes in domestic emissions (e.g., for EU = ΔNA/ΔEU). NA contributions range from 1 to 3 percent for surface PM and 5 to 17 percent for column sulfate, with the contributions to EU>SA>EA. The increase in import sensitivity from surface concentrations to column loadings indicates the substantial role of aerosol transport above the planetary boundary layer in the long-range transport of PM. The column aerosol loadings play an important role in radiative forcing. The fact that the import sensitivity of total column aerosol load is significantly larger than that for the surface concentrations means that regional aerosol radiative forcing and climate change are more greatly affected by distant sources than are air quality goals.

The HTAP comparison results provide insights into how the uncertainties in the models impact the source-receptor relationships. The chemical, removal, and emission processes all affect the lifetime of the species in the atmosphere. The impact of one region on another depends on the transport pathways, their length, and the intensity of the removal processes that the PM experiences along its way. The model-to-model diversity of sulfate lifetimes in absolute terms for a given region is considerable (a factor of four), and reflects differences in the process level model formulations. In general, models that predict longer lifetimes predict larger contributions from distant sources. In addition, emissions from North America and East Asia appear to be slightly more efficiently removed than those from Europe and South Asia. These results suggest the need to develop observational techniques to constrain the lifetime of PM in the atmosphere, which would provide a strong metric for evaluating and improving models.

The use of the PM source attribution information in the assessment of impacts is just beginning. For example, Liu et al. (2009) use tagging to identify intercontinental source-receptor relationships, and estimate that approximately 90,000 premature deaths (for adults age 30 and up) globally and 1,100 in North America are associated with exposure to nondust $PM_{2.5}$ of foreign origin. It is worth noting that this number of premature mortalities in North America is comparable to the reduction in premature mortalities expected to result from tightening the U.S. 8-hr O_3 standard from 84 ppbv to 75 ppbv.

FUTURE PROJECTIONS OF FACTORS INFLUENCING PM SURFACE CONCENTRATIONS

A 20 percent increase in emissions in Europe, East Asia, and South Asia is estimated to change the annual $PM_{2.5}$ concentrations at the surface in North America by ~ 0.02 $\mu g\ m^{-3}$ (HTAP-TF, 2007). This import contribution amounts to 2 percent of the proposed annual average $PM_{2.5}$ PSD increment (1 $\mu g\ m^{-3}$), and 50 percent of the proposed significant impact level, 0.04 $\mu g\ m^{-3}$, in Class I areas. The significant impact level is the level of ambient impact that is considered to represent a significant contribution to nonattainment. Larger increments are allowed in other areas, thus implying the biggest relative impact from imported emissions in national parks and wilderness areas. However, the estimated ~ 0.02 $\mu g\ m^{-3}$ represents 10 percent of the allowed interstate impact under CAIR. The emissions of major PM components and precursors (BC and SO_x) are estimated to have increased in Asia by ~ 50 percent from 2000 to 2006 (Zhang et al., 2009a), over two times the 20 percent assumed in the modeling experiment. Regional and seasonal contributions can be much higher. Although the most significant impacts on U.S. air quality from long-range transport of PM are likely to be related to compliance with the regional haze rule, the impacts extend beyond Class I areas.

Although changes in baseline PM concentrations currently appear insignificant in comparison with local contributions to NAAQS exceedances in urban areas, as standards tighten a rising baseline would represent an increasing percentage of the total that cannot be controlled by domestic regulations. Furthermore, imported PM may represent a disproportionately higher fraction of specific aerosol components that may be targeted in future regulations. Thus, shifting future priorities may enhance the relative importance of non-U.S. contributions to PM. It should also be noted that changes in emissions from foreign sources may change the magnitude of extreme events, in addition to the baseline changes mentioned above.

In Table 3.1 we provide rough estimates of future emission changes based on the knowledge available today. The remainder of this section describes the sources of the values in the table. Large uncertainties in these estimates can be expected, due to imperfect understanding about the trajectories of economy and technology, and about the responses of forests and deserts. The values here are provided only to evaluate whether future changes in intercontinental transport could be expected to significantly affect PM budgets in the United States. Most models that examine changes in future emissions due to anthropogenic global change compare present-day conditions with a doubled-CO_2 climate (expected to occur ~ 2050) or with conditions in 2100. This study is focusing on a shorter time horizon, roughly to 2030; to approximate the response in 2030 we use half the change projected for 2050 (doubled CO_2), or one-quarter of the change projected for 2100 (recognizing that this approach may miss some non-linearities in the actual response). We rely on model results summarized by the recent Intergovernmental Panel on Climate Change report (IPCC, 2007) when possible.

TABLE 3.1 Possible Future Changes in Emissions of PM (see text for explanation of how numbers were derived).

Source Region	Receptor Region	Percent Change by 2030
Energy and Economy (Change in Baseline)		
* East Asia	Western U.S.	– 40 to + 50
* Mexico and Central America	SE U.S.	– 80 to – 50
† Northeast U.S.	Canada, Mexico, Europe, Arctic	– 80 to – 40
* Shipping	Coastal U.S.	– 40 to + 55
Dust (Change in Baseline and Episodes)		
* East Asia	Western U.S.	– 15 to + 95
* North Africa	SE and NE U.S.	
Open Fire (Change in Episodes)		
* Mexico and Central America	SE and SW U.S.	
* Western Canada	Midwestern U.S.	+10 to +30
† Northeast U.S.	Arctic	

* Import to United States.
† Export from United States.

Energy and Economy   We use the sum of the increases in sulfur dioxide, black carbon, nitrate, and organic carbon emissions to estimate PM increases for 2030. Our range of estimates is based on the maximum feasible reduction and current legislation scenarios produced by IIASA (Cofala et al., 2007). These changes are −45 to +50 percent for East Asia, −80 to −50 percent for Central America, and −60 to −80 percent for the United States. A wider range of economic and technological scenarios has been explored for shipping emissions (Eyring et al., 2005), and that range is also given in the table. The contribution of secondary organic particulate matter is not included in this analysis, but estimates of future increases are less than 10 percent by 2030, based on 2100 estimates (Heald et al., 2008).

Dust   Desert dust is mobilized by wind from exposed surfaces. Dust emissions are affected by the nature of the land surface, which changes due to human activity and vegetation (McConnell et al., 2007; Neff et al., 2008), and by surface windspeed (Tegen, 2003; Engelstaedter and Washington, 2007). These governing factors result in significant annual variability (Prospero and Lamb, 2003; Qian et al., 2004) and sensitivity to climate, including emission damping by increased vegetation in a higher-CO_2 world (Mahowald and Luo, 2003). IPCC (2007, Section 7.5.1.1) provides a range of changes by 2100, from −60 to 380 percent, and we use one-quarter of this value.

Open Fires   Open fires occur because of deforestation, natural causes such as lightning strikes, prescribed fires for land management, and seasonal fires for agricultural land clearing. In our estimates of future changes we neglect two of these: deforestation, which is a minor factor in most source regions, and prescribed fires, which depend on management policies. A warmer climate is expected to increase fire frequency and quantity (IPCC, 2007, Section 7.3.3.1; Spracklen et al., 2007). This change is attributable to droughts and to other changes in climate regimes for both tropical areas and boreal forests. We infer, from estimates of changes in Canadian fires by 2100 (Flannigan et al., 2005; Girardin and Mudelsee, 2008), that an increase of 10-30 percent could be expected by 2030. However, many of these estimates do not account for all feedbacks, such as changes in vegetation or ignition frequency. Further, the burned area of Canadian forest fires has already doubled between the 1970s and 1990s (Kasischke and Turetsky, 2006). Thus, the value in Table 3.1 could be an underestimate.

Transport   Pollutants arriving in either the United States or regions affected by the United States, may be altered due to future synoptic-scale dynamics, convection and transport at high altitudes, or climate modes such as El Niño or the North Atlantic Oscillation (HTAP-TF, 2007). If the average climate changes, we may also expect differences in reaction rates, such as those leading to secondary organic particulate matter, and removal rates, such as rainout. While specific studies focused on the impact of climate change on the long-range transport of pollutants are just beginning, a number of robust changes in transport patterns under climate change scenarios have emerged from model, theoretical, and observational studies (HTAP-TF, 2007). These studies suggest less frequent large-scale venting of the boundary layer, which implies less export of pollution from the boundary layer. However, once exported from a source region, the extent of long-range transport depends upon the lifetime of the pollutant. For the case of particulate matter the lifetimes are heavily dependent on the wet removal processes (i.e., the lifetimes decrease as wet removal rates increase). Changes in cloud properties and precipitation patterns and quantity resulting from climate change will affect the transport distances of PM. Thus, if PM lifetimes are decreased, export from source regions may remain unchanged or even decrease despite increasing future emissions. We also know that pollution export is dependent upon location (HTAP-TF, 2007), with export from Southeast Asia being more efficient than from North America and export from Europe least efficient. This is due to the amount of convective activity over oceans (Stohl et al., 2002). Future emission increases are projected to be shifted toward the tropics (IPCC, 2007), and the efficiency of export from those regions experiencing increases need to be considered in order to evaluate their impact on long-range transport.

KEY FINDINGS AND RECOMMENDATIONS

Question: What do we know about the current import and export of PM?

Finding. Satellite, aircraft, and ground-based observations demonstrate that PM can undergo long-range transport. Episodic events associated with biomass burning plumes, dust storms, and fast transport of industrial pollution are most easily identifiable in observations from rural areas and high altitudes, where they represent large excursions above typical ambient concentrations. Some excursions in PM concentrations clearly attributable to long-range transport episodes have been recorded at the surface in urban areas, but in general, the contributions from long-range transport in such regions are difficult to isolate from PM of U.S. origin.

Finding. It remains difficult to quantify from observations alone the frequencies and magnitudes of contributions from persistent long-range transport of PM. Such transport occurs routinely, but presumably brings much lower concentrations of PM than do episodic events, making it difficult to quantitatively distinguish the long-range transport signal from that due to local and regional PM sources.

Finding. Ensemble studies of chemical transport models estimate that PM sources in Europe, South Asia, and East Asia contribute on average 0.05–0.15 µg m^{-3} to the continental U.S. annual mean surface PM$_{2.5}$ concentration (obtained by scaling the 20 percent perturbation results by a factor of 5 to account for 100 percent of the foreign influence, ignoring any nonlinear effects). Larger contributions can occur for particular seasons and geographical regions.

Finding. Chemical transport models can be used to estimate long-range transport contributions to atmospheric PM concentrations, as well as the contributions to deposition of sulfate, nitrate, metals, Hg, and POPs. However, large uncertainties in model estimates result from several factors, including lack of observational con straints, particularly in the free troposphere where much of the transport occurs; poorly constrained or unknown emissions of some primary particles and the emissions and conversion of PM precursors (especially for secondary organic PM); and poorly con-strained wet and dry PM removal processes, as reflected in the large differences in PM lifetimes among models.

Finding. There are growing capabilities for observing total column amounts of PM from ground-based and satellite observations, and this information is being used to detect global and regional column trends. Yet the linkages between column PM amounts and surface PM levels are complicated and not well understood; for example, there is little information about how PM aloft is mixed down to the surface. This limits our ability to quantify the effect of persistent transport on surface concentrations based on column observations. There is, however, ongoing work to improve our ability to model these relationships between column and surface PM.

Finding. There is not enough observational evidence to demon-strate clear trends in average baseline levels of surface PM on a global or hemispheric scale. Because of the relatively short lifetime of PM, observations must be gathered over long time periods (around a decade at least) before trends can be distinguished from short-term variability.

Recommendations. Improve our ability to characterize the import and export of PM by implementing

- continuous aerosol observatories in key locations, to be defined as part of a comprehensive measurement strategy, including strategically located marine- and mountaintop-based sites. Key observational variables include those that can help with identifying source fingerprints (e.g., speciated and size-resolved aerosols) to permit quantification of the associated PM fractions with high time resolution. A minimum requirement is one sample per 24 hours, but shorter time resolution will yield even greater capability to distinguish between local and long-range transport. Long-term observations are needed to encompass the statistical variability of natural processes, as well as to separate trends in emissions from variability in transport and removal.
- long-term, time-resolved data collection, both inside and outside urban areas, combined with meteorological modeling of horizontal and vertical transport. Such data are needed to elucidate the impact of long-range transport in regions that are near air quality standard compliance limits, where the contribution of long-range transport is generally most difficult to detect.
- improvement and application of methods for quantifying import and export of PM using space-based observations. A particular focus should be on observations and methods that can be used to constrain vertical distributions (e.g., space-based and ground-based lidars) and those that can elucidate the relationships between column amounts of PM and surface PM concentrations.
- advancements in modeling capabilities that include merging bottom-up with top-down emission estimates; validating combined emission and model systems with in situ and remotely sensed observations; constraining PM column amounts, including speciated PM when possible; and refining and applying fingerprinting techniques in observations to aid in verification of modeled source attributions. Modeling improvements that are particularly important in evaluation of long-range transport are the simulation of dust emissions, fire emissions, and production of secondary organic PM.

Question: What are potential implications of long-range PM transport for meeting environmental goals?

Finding. With the exception of occasional extreme episodes, most often attributable to dust and wildfires, long-range transport of PM is estimated to represent a negligible contribution to PM NAAQS exceedances in most regions in the United States.

Finding. Increases in surface PM from episodic or persistent transport are significant relative to other sources in places such as the Arctic and portions of the western United States, which do not have large local pollution sources. In particular, strategies to achieve compliance with the regional haze rule across the United States may be complicated by imported PM because existing U.S. regulations prohibit increments in annual mean $PM_{2.5}$ concentrations above 1 µg m^{-3} in Class I areas, and even tighter limits on incremental contributions may be imposed in future regulations.

Finding. Some initial studies have begun to estimate the premature mortalities attributable to long-range transport of PM, but they are very preliminary at this time due to the large uncertainties in estimates of imported and exported PM. These estimates are not only limited by uncertainties in quantitative estimates of the imported fraction of PM but also by lack of a mechanistic understanding of how the individual components of PM are linked to health. Current estimates indicate that domestic sources of PM are by far the larger risk to human health. Any import of PM from distant sources could add to the health burden, because there is thought to be no threshold for PM health impacts (and impacts rise linearly with increasing concentration).

Finding. Both longer-term sustained and episodic increases in column PM across the United States attributable to long-range transport may represent potentially significant perturbations to the regional radiative balance.

Finding. The ecological and agricultural impacts of long-range transport of PM have not been evaluated.

Recommendation. Improve abilities to estimate the potential implications of non-U.S. PM sources for the attainment of U.S. environmental goals by:

•   improving quantification of imported and exported PM fractions, which in turn requires coordinated modeling and observational strategies (see next question);
•   developing the capability to quantify the trends in PM surface concentrations and to attribute these trends to emissions from

specific sources. Because baseline PM concentrations are so low and variable, questions regarding trends in this baseline can be answered only through a long-term strategy that integrates observations and models. Baseline and background changes tend to be regional (e.g., Northern Hemisphere; Arctic, Pacific, and Atlantic regions), thus requiring corresponding regional strategies;

• reevaluating the health burden resulting from imported PM, as better mechanistic understanding of PM-health links becomes available;

• assessing the other impacts of imported PM, such as adverse consequences for ecological and agricultural systems, and consequences for regional and global radiative forcing.

Question: How might factors influencing the relative importance of U.S. and foreign contributions to PM change in the future?

Finding. A warming climate is expected to lead to shifts in emissions of PM associated with the natural environment and changes in the meteorology that affects pollution dispersion. Hypothesized impacts include:

• increasing annual numbers of forest fires, which in turn would increase the frequency of extreme PM concentration events from long-range transport of fire emissions;

• changes in dust particle sources (although even the sign of increase or decrease is currently uncertain);

• changes (often reductions) in the rate at which surface and boundary layer air is vented to the upper atmosphere;

• complex changes in cloud amount and lifetime and precipitation patterns, in turn affecting PM removal rates and spatial patterns of deposition;

• changes in PM precursor emissions, particularly biogenic gases emitted from vegetation.

These changes would all have implications for how much PM (and precursors) is emitted from various source regions, how much of the emitted PM escapes from source areas, and the fraction of PM that survives transport to affect areas downwind. At present, it is not possible to predict the net impacts.

Finding. In the short term (e.g., next 10–20 years), emissions from developing countries will likely continue to increase. In the longer term it is not possible to predict how future emissions will change and how these will in turn affect long-range transport of PM. As global economic and industrial activities expand, greater emissions of PM and precursor species might be expected, with projected growth varying widely, depending on economic scenario assumptions. The result could be a higher baseline of imported PM; however, introduction of standards, and cleaner technologies to address national standards moving toward WHO guidelines, may slow or reverse this trend.

Finding. A better understanding of the health effects of PM may lead to different or more stringent standards in the United States and elsewhere. For example, revised regulations may not focus solely on mandating lower acceptable concentrations of $PM_{2.5}$ but may also (or instead) focus on particular fractions or characteristics of PM, such as number concentration, the presence of toxic metals, or the particles' oxidative state. Identification of the PM characteristics responsible for health outcomes will also allow monitoring, modeling, and apportionment activities to be tailored toward those characteristics, which will in turn help advance our understanding of the true significance of long-range transport of PM.

Recommendation. Increase our ability to forecast the future significance of imported PM by:

• increasing understanding of how future climate will affect patterns of dust mobilization, fire frequency and intensity, and biogenic emissions;
• understanding how regional perturbations in aerosols affect climate;
• developing multipollutant emission inventories with high spatial resolution that reflect economic growth, technological change, and incoming regulations;
• improving the forecasting of changes in atmospheric dynamics that affect transport of pollution plumes, particularly boundary layer venting and precipitation, under changing climates;
• designing effective observational networks that enable rigorous analysis of PM composition, trends, and other factors that relate to long-range transport.

Additional information will be needed to pinpoint the PM characteristics and monitoring strategies that are most useful for understanding impacts on human health.

4

Mercury

ATMOSPHERIC MERCURY PRIMER

The element mercury (Hg) is a unique metal that is liquid at ambient conditions, and easily volatilized to the atmosphere, where it is distributed on a global scale. This environmental contaminant emitted from any source has the potential for long-range transport, interaction with and assimilation by terrestrial and aquatic surfaces, and a ubiquitous presence in all environmental media. The dominant form of Hg in the atmosphere is elemental (Hg (0)) and concentrations are typically in the range of 1 to 2 ng m^{-3}. Other forms, including divalent Hg (II) and monovalent Hg (I) gaseous compounds and particulate bound Hg (Hg$_p$), are usually less than 5 percent of the total atmospheric burden (Schroeder and Munthe, 1998; Valente et al., 2007). These small concentrations make measurement of atmospheric Hg challenging. Gaseous divalent forms referred to here as reactive gaseous mercury (RGM) are thought to include compounds such as HgCl$_2$, Hg (OH)$_2$, HgBr$_2$, and HgO (Lin and Pehkonen, 1999). RGM may be a primary or secondary air pollutant. In cloud water, Hg can be dissolved as Hg (0) or Hg (II) (Lin and Pehkonen, 1999).

The major forms that concern human exposure are Hg (0), and compounds containing methyl mercury (methyl Hg). These have different health effects and exposure pathways with Hg (0) being most harmful when inhaled and methyl Hg when ingested. Hg (0) affects the nervous system, kidneys, and lungs (Clarkson, 2002); methyl Hg is a neurotoxin, teratogen, and linked with cardiovascular problems (Clarkson, 2002; Mozaffarian and Rimm, 2006). Methyl Hg is produced in ecosystems from inorganic mercury derived from the atmosphere or terrestrial landscape (Munthe et al., 2007). Although this form has been measured in rain, the concentrations are extremely low, and the atmosphere is not thought to be a direct source to ecosystems (Sakata and Marumoto, 2005; Hammerschmidt et al., 2007).

Hg is released to the atmosphere by both natural and anthropogenic sources. Natural sources emit almost exclusively Hg (0) (Bagnato et al., 2007; Gustin et al., 2008) ; anthropogenic sources emit varying combinations of Hg (0), RGM, and Hg_p (Pacyna et al., 2003b, 2006a). Atmospheric RGM may also be generated by reactions of Hg (0) with oxidants such as O_3, OH, and reactive halogens such as Cl, Cl_2, Br, and BrO (Lin et al., 2006; Hynes et al., 2008; Ariya et al., 2009). These reactions, which may occur in the surface boundary layer, the free troposphere, and stratosphere, and are poorly understood.

To evaluate the potential for long-range transport of Hg once emitted from a natural or anthropogenic source, the atmospheric lifetime (mean time in the air before being removed) of the different forms needs to be considered. Gaseous Hg (0) has an estimated atmospheric lifetime of months to more than a year (Lindberg et al., 2007), thus a molecule of Hg in a fish may have its origin from a source far removed. Episodic events of elevated air Hg concentrations recorded at the Mt. Bachelor Observatory in central Oregon (Jaffe et al., 2005a, Figure 1.3; Weiss-Penzias et al., 2006) and in aircraft measurements (Friedli et al., 2003; Swartzendruber et al., 2008) have been linked to air masses passing over Asia. The form of Hg transported in these events is predominantly Hg (0). Elevated air Hg concentrations have also been measured in plumes associated with fires and industrial sources (Edgerton et al., 2006; Ebinghaus, 2008; Finley et al., 2009). In contrast RGM and Hg_p are water soluble and have high deposition velocities, resulting in efficient removal from the atmosphere by dry and wet deposition (Schroeder and Munthe, 1998).

Field and laboratory data have shown that Hg (0) is recycled between the air and terrestrial and aquatic surfaces, and this can occur over the course of a day (Gustin et al., 2008). Limited field studies using stable isotopes and laboratory experiments have indicated that 5 to 40 percent of Hg added to an ecosystem as $HgCl_2$ is released in the short term (Hintelmann and Evans, 1997; Hintelmann et al., 2001, 2002; Lindberg et al., 2002; Amyot et al., 2004; Ericksen et al., 2005; Xin et al., 2007). This recycling of Hg between the air and surfaces results in long-term availability of Hg to ecosystems and complicates source attribution.

Human and Ecosystem Health   The major human and ecosystem health threat associated with long-range transport of Hg arises from the fact that Hg in the air is deposited to watersheds where it may enter aquatic systems and be converted to methyl Hg (Munthe et al., 2007). Humans are exposed to this neurotoxin primarily by fish consumption (~ 95 percent of the Hg in fish muscle tissue is methyl Hg [Bloom, 1992]). Because Hg is globally distributed, fish in remote regions may be impacted by any source. In the United States, 48 states have Hg consumption advisories, (EPA, 2006b) and many are associated with water bodies located in areas with no apparent land-based Hg contamination or anthropogenic Hg source. Consumption advisories are set by the states; some states apply the 0.3 μg g^{-1} (0.3 ppm) EPA limit while others apply the 1 ppm FDA limit.

Since methyl Hg is bioconcentrated in organisms and biomagnified in aquatic food webs, large fish and those with high trophic stature (i.e., bass, walleye, and perch) tend to have higher concentrations. Thus, marine and freshwater advisories often target specific fish species and are size based. Of the marine fish, shark, tuna, swordfish, and tilefish have been shown to have elevated concentrations of Hg (Burger and Gochfeld, 2004; Chen et al., 2008). High blood levels of Hg have been reported for those eating large quantities of marine fish high in the food web (Hightower and Moore, 2003).

An NRC committee (NRC, 2000) concluded that the risk of adverse effects from current methyl Hg exposures to the majority of the U.S. population is relatively low. However, since the developing human nervous system is sensitive to methyl Hg, young children and children of women who consume fish during pregnancy are at risk (IPCS-WHO, 1990; NRC, 2000; Clarkson et al., 2003). The NRC committee recommended, as have others (Sakamoto et al., 2005; Mozaffarian and Rimm, 2006), that since fish are an important food resource benefiting human health, the long-term goal should be reduction of methyl Hg concentrations in fish rather than replacement of fish in the diet by other foods.

Humans are not the only organism at risk due to Hg exposures. Scheuhammer et al. (2007) concluded that dietary methyl Hg exposures at environmentally relevant concentrations may cause behavioral, neurochemical, hormonal, and reproductive effects in wild animals. Similar results were reported for Hg-exposed fish in national parks of the western United States (Landers et al., 2008). Other research has suggested that marine predators such as sharks, seabirds, seals and walruses, may also be at risk (Kemper et al., 1994; Braune et al., 2006; García-Hernández et al., 2007).

Policy and Regulatory Context   Current U.S. regulations for Hg are limited to the listing as a hazardous air pollutant in the 1990 Clean Air Act amendments. In March 2005, the EPA tried to remove electricity-generating utilities from the listing, instead allowing for development of a cap-and-trade program called the Clean Air Mercury Rule (CAMR). This Rule was vacated by the DC Circuit Court in February 2008 and the future of this legislation is unclear (Milford and Pienciak, 2009).

Discussions are ongoing within United Nations Environmental Program (UNEP), the European Union, and U.S.-Mexico-Canada about how to deal with this transboundary pollutant. Because Hg is a global pollutant, reducing Hg deposition to terrestrial and aquatic ecosystems ultimately requires an international protocol. At its session in February 2007 the UNEP Governing Council concluded that efforts to reduce risks from mercury were not sufficient to address the global challenges posed by mercury; and they concluded that further long-term international action is required (http://www.chem.unep.ch/).

ATMOSPHERIC MERCURY CONCENTRATIONS

Atmospheric Hg concentrations measured in remote locations are relatively consistent over space and time with values of 1.5 ± 0.2 ng/m^3 reported for the Northern Hemisphere and ~ 1.2 ng/m^3 for the Southern Hemisphere (Slemr et al., 2003; Lindberg et al., 2007). Fairly constant concentrations have been reported for the past 10 to 15 years, based on data collected during ship cruises and at Mace Head, Ireland. The ubiquitous nature of Hg in the atmosphere results in this reservoir being described as a "global pool" that is a mixture of Hg emitted from all sources (Lindberg et al., 2007). Higher atmospheric Hg concentrations are reported for areas directly affected by anthropogenic and natural sources (Ebinghaus, 2008 and references therein). Reductions in air Hg concentrations on a regional scale have been reported for areas where anthropogenic emissions have been reduced (Kellerhals et al., 2003; Wängberg et al., 2007). However there are limited long-term air concentration datasets and significant gaps in the spatial and temporal coverage (Keeler et al., 2009). Since Hg is a global pollutant, understanding Hg is critical for assessing potential sources to the United States.

RGM and Hg$_p$ concentrations in air are typically in the tens of pg m^{-3}. Long-term trend data are not available for these forms, because automated methods to measure these forms have only been available during the past ten or so years, and regular monitoring has been done at only a few locations. In remote continental areas RGM concentrations often peak during the day and decline to zero at night (see references listed in Ebinghaus, 2008). RGM and Hg$_p$ are emitted directly by specific point sources. RGM may also be produced by oxidation of Hg (0) to RGM. While the exact nature of this oxidation is not understood, there is evidence that this process occurs in the Arctic boundary layer (Dommergue et al., 2008; Steffen et al., 2008), marine boundary layer (Laurier et al., 2003; Hedgecock and Pirrone, 2004; Sprovieri et al., 2008), surface boundary layer (Weiss-Penzias et al., 2003; Liu et al., 2007; Peterson and Gustin, 2008), free troposphere (Swartzendruber et al., 2006; Sillman et al., 2007), and possibly in the stratosphere (Talbot et al., 2007; Slemr et al., 2009). A few studies have reported elevated RGM at high altitudes using mountain top and aircraft sampling platforms (Landis et al., 2005; Swartzendruber et al., 2006; Sillman et al., 2007). Subsidence of the air from the free troposphere and convective mixing may result in delivery of Hg, and specifically RGM, from the global pool to the surface. Recent field and modeling work has suggested that this is important in Nevada (Selin et al., 2007; Weiss-Penzias et al., 2009) and in the southeastern United States (Guentzel et al., 2001; Selin et al., 2008). The process of oxidation of Hg (0) in the free troposphere and subsequent transport to the surface could be an important means by which Hg is deposited from the global pool.

SOURCES AND SINKS OF ATMOSPHERIC MERCURY

A recent UNEP initiative (Pirrone and Mason, 2009a) estimated current annual emissions of Hg to the atmosphere as ~ 8000 Mg/yr, with approximately one-third directly derived from anthropogenic sources. Natural sources are thought to contribute an additional one-third of the total surface-to-air flux each year. The remainder of the annual Hg input to the atmosphere is that part previously emitted from natural and anthropogenic sources designated as "legacy Hg" but still being cycled between surface reservoirs and the atmosphere. Because Hg is not degraded, once emitted from a source it may become a component of the legacy pool until some process (geophysical, geochemical, biological) results in long-term removal and sequestration. Figure 4.1 illustrates the important fluxes and reservoirs associated with the Hg biogeochemical cycle. The atmosphere itself is a small reservoir.

The major anthropogenic sources of atmospheric Hg are coal combustion, waste incineration, chlor-alkali plants, and metal processing (Table 4.1). However emission inventories are evolving with new sources being identified and estimates for specific sources and developing countries having a high degree of uncertainty. The recent UNEP report and resulting book (Pirrone and Mason, 2009a,b) provide a summary of the current estimates of contributions from anthropogenic sources in developed and developing countries. Inventories for latter are incomplete and have uncertainties on the order of 35 to 50 percent (Lindberg et al., 2007). Based on the UNEP report, China and India are estimated to contribute 42 and 20 percent, respectively, of current global anthropogenic emissions, while North America contributes 9 percent (Table 4.1). As developing countries grow economically their fossil fuel combustion is projected to increase. Wu et al. (2006) estimated an increase in China's Hg emissions of 3 percent

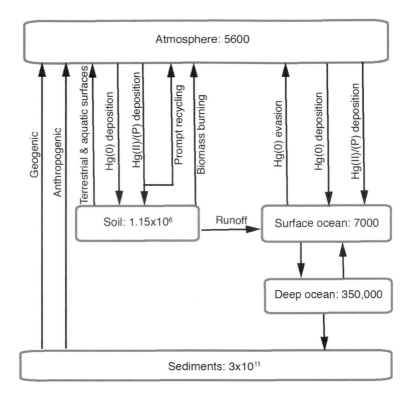

FIGURE 4.1 Major reservoirs and fluxes in the global Hg biogeochemical cycle, adapted from Selin et al. (2008). See text for discussion of values applied, as well as Pirrone and Mason (2009a,b) for additional information. Mercury reservoirs are shown in closed boxes in units of Mg.

per year from 1995 to 2003, primarily due to increasing coal combustion. Streets et al. (2009) suggested that the change of global anthropogenic Hg emissions may range anywhere from –4 to +96 percent by 2050, depending on the future of coal-fired utility releases. Estimated individual source contributions also have significant uncertainty on the order of ±25 percent for stationary fossil fuel combustion, ±30 percent for non ferrous metal production and industrial sources, and ±50 percent for gold production (Pacyna et al., 2006a; Lindberg et al., 2007).

Geogenic emissions of Hg to the atmosphere occur from areas with ongoing volcanic and geothermal activity and from substrates with Hg concentrations enriched by past geologic activity above the estimated natural background of < 100 ppb (Wedepohl, 1995). Estimated emissions of Hg from naturally enriched substrates range from 500 to 1500 Mg/yr (Gustin

TABLE 4.1 Global Anthropogenic Emissions by Countries and Regions

	Stationary combustion	Non-ferrous metal production	Pig iron and steel production	Cement production	Caustic soda production	Mercury production[a]	Gold production	Waste disposal	Other	Total	Reference year	Reference
South Africa	11.8	0.4	1.3	3.8			2.3	1.4	0.9	21.8	2007	1
China	268.0	203.3	8.9	35.0		27.5	133.4	10.4	11.3	697.8	2003	2
India	120.9	12.1	2.9		106.0		0.5	77.4	8.0	327.8	2004	3
Australia	2.2	11.6	0.8	0.3			0.3	0.2	0.6	16.0	2005	4
Europe	113.9	15.4	12.5	30.2	40.4			11.6	15.3	239.3	2000	5
Russia	18.7	7.2	2.6	1.6	1.2		3.3	3.6	0.3	38.5	2001	6
North America	65.2	34.7	12.8	7.0	10.3			13.0	1.7	144.7	2003	7
South America	31.0	25.4	1.4	6.5	5.0	22.8	67.6			159.7	2000	8
Total	631.7	310.1	43.2	84.4	162.9	50.3	207.4	117.6	38.1	1645.6		

References: (1) Leaner et al., 2008; (2) Streets et al., 2008; (3) Mukherjee et al., 2008; (4) Nelson, 2007; (5) Pacyna et al., 2006b; (6) ACAP, 2005a, b; (7) Pirrone, 2008; (8) Pacyna et al., 2006a.

NOTE: Mercury emissions from artisanal mining here reported have been estimated by Telmer and Veiga (2008). Global mercury emission account for 330 Mg yr⁻¹ (Africa 24 Mg yr⁻¹, Asia 208 Mg yr⁻¹, South America 68 Mg yr⁻¹).

SOURCE: Pirrone, 2008.

et al., 2008). The range in estimated releases from volcanic systems is large (1 to ~ 700 Mg/yr, Gustin and Lindberg, 2005). The only global estimate for geothermal emissions indicates a source of 60 Mg/yr (Varekamp and Buseck, 1984). Both volcanic and geothermal emissions will vary temporally depending on activity.

Wildfires (biomass burning) contribute Hg to the air, with estimates of global emissions ranging from 200 to 1000 Mg/yr (Brunke et al., 2001; Friedli et al., 2001; Weiss-Penzias et al., 2007). While most of the release from wildfires is in the form of Hg (0), particulate-bound Hg may be a component of the Hg released (Wiedinmyer and Friedli, 2007; Finley et al., 2009). Mercury derived from this source is largely legacy Hg (Wiedinmyer and Friedli, 2007; Gustin et al., 2008). Some component of the Hg emitted during fires may be derived from soils; the importance is unclear and most likely related to fire intensity (Engle et al., 2006; Biswas et al., 2007).

Hg emitted from terrestrial and aquatic surfaces (not enriched by geologic processes) and the oceans is natural and anthropogenic in origin. Most of the Hg emitted from the former can be accounted for by inputs from the atmosphere (Gustin et al., 2008; Hartman et al., 2009). All forms of atmospheric Hg may be deposited to soils; the potential for long-term sequestration after deposition is not clear. Recent work has suggested that terrestrial ecosystems are a net sink for atmospheric Hg (Hartman et al., 2009). The overall role of the oceans as a sink or source for atmospheric Hg, and the relative contributions of natural and legacy Hg are debated. Earlier work suggested that atmospheric inputs to the ocean were roughly equal to ocean emissions (Mason et al., 1994), but more recent global mass balances have suggested the oceans are a net sink of 500 to 2500 Mg/yr (Lamborg et al., 2002b; Mason and Sheu, 2002; Sunderland and Mason, 2007). A recent model-based study suggested that the majority of the oceanic emissions are legacy Hg (Strode et al., 2007).

Ocean sediments, plant foliage, and polar regions are sinks for atmospheric Hg. The latter are considered to be a small sink where Hg (0) is oxidized to RGM by atmospheric bromine compounds (< 100 Mg/yr) (Dastoor et al., 2008; Outridge et al., 2008). After formation, RGM is deposited to snow, with a portion remaining in the ecosystem and the rest emitted back to the atmosphere (Poulain et al., 2004; Kirk et al., 2006). Plant foliage is a significant sink for atmospheric Hg (0) (Gustin and Lindberg, 2005) and plant litter is a significant reservoir (Grigal, 2003). Ultimate removal of Hg from the atmosphere-ocean-land system occurs through settling of particulate Hg to the deep ocean where it is incorporated into sediments (Strode et al., 2007; Sunderland and Mason, 2007). Model studies suggest that the time scale for this removal is thousands of years (Selin et al., 2008).

ATMOSPHERIC MERCURY DEPOSITION

The potential for Hg introduction to aquatic food webs by way of wet and dry deposition is a primary concern. Wet deposition measurements of Hg are routinely made in the United States as part of the National Mercury Deposition Network (MDN) (http://nadp.sws.uiuc.edu/mdn/). Observations have been made for up to twelve years at some sites. It should be noted that most MDN sites were chosen to sample regional background air, and in most cases are removed from local sources. Spatial patterns in wet deposition measured as part of this network are similar from year to year, and recent work has suggested slight declines in deposition at specific locations in the eastern United States (Butler et al., 2007; Prestbo and Gay, 2009). Based on the maps available from the MDN network, the highest wet deposition rates occur in Florida and along the Gulf Coast (up to 16 $\mu g/m^2$ yr); and the eastern states have higher wet deposition inputs (7 to 10 $\mu g/m^2$ yr) than those in the west (1.2 to 5 $\mu g/m^2$ yr). Several studies have reported that wet deposition is enhanced directly around point sources (Dvonch et al., 1999; Munthe et al., 2001; Keeler et al., 2006; Wängberg et al., 2007). White et al. (2009) reported that 42 percent of the mercury in summertime wet deposition at the Steubenville, Ohio, MDN site, an area with a high density of coal-fired utilities, could be linked to local point source emissions.

There is no standard method for measuring Hg dry deposition, and this process is poorly understood. Spatial and temporal variability in dry deposition is likely to be large since it varies inversely with precipitation. In a few field studies dry deposition was shown to be of similar magnitude as wet deposition (Lamborg et al., 2002a; Caldwell et al., 2006; Lyman et al., 2007). Model studies suggest that for the United States as a whole, dry deposition is greater than wet deposition with significant spatial variability (Selin et al., 2008). A recent study that applied surrogate surfaces for measuring dry deposition found wet:dry inputs in Nevada to be 1:4.5, while in the southeastern United States this was 6:1 (Lyman et al., 2009).

Indirect methods that have been applied to estimate net Hg atmospheric deposition include the use of sediment cores (Swain et al., 1992; Lucotte et al., 1995; Landers et al., 1998, 2008; Bindler et al., 2001; Lamborg et al., 2002a; Yang et al., 2002; Fitzgerald et al., 2005), ice cores (Schuster et al., 2002), and peat bog profiles (Biester et al., 2007 and references therein). It is thought that sediment cores are the more reliable indicator of historic deposition (Biester et al., 2007). Data indicate that Hg deposition has increased significantly over the past 150 years in both remote and industrialized areas. In remote regions increases in deposition of two to five times above preindustrial rates are reported (Lucotte et al., 1995; Engstrom and Swain, 1997; Fitzgerald et al., 1998; Lamborg et al., 2002a). The measured increase occurs concurrently with industrialization, suggesting that anthropogenic sources are responsible (Biester et al., 2007). Increases in deposition observed in remote regions indicate that Hg in the global atmosphere pool is an important component.

Although sediment cores clearly show an increase in deposition since the industrial revolution, interpretation of the sediment core data may be more complex than originally believed. Mercury in some cores has been found to be correlated with phytoplankton (Outridge et al., 2005) and organic carbon (Landers et al., 2008). Since Hg is readily bound to carbon, changes in watershed characteristics and climate that influence the available carbon may affect the ability of the sediments to permanently sequester Hg.

RELATIONSHIP BETWEEN MERCURY DEPOSITION AND ECOSYSTEM IMPACTS

For both freshwater and marine ecosystems the relationship between Hg deposition and Hg concentrations in fish is not straightforward because the properties of individual ecosystems will affect the efficiency with which Hg is converted into methyl Hg. It is believed that increased Hg deposition to aquatic ecosystems leads to increases in methyl Hg in fish, but there is limited quantitative data to test this hypothesis (Munthe et al., 2007).

One study that investigated the potential for fish uptake of atmospherically deposited Hg (the METAALICUS project) (Harris et al., 2007) demonstrated that increased Hg concentrations in aquatic organisms were clearly linked with the direct loading of a specific stable isotope of Hg in precipitation to the lake surface. However, < 1 percent of the stable isotope added to the land and vegetation surrounding the lake was found in the lake ecosystem after three years.

Other studies have suggested a direct link between atmospheric inputs and biota Hg concentrations. For example, Hammerschmidt and Fitzgerald (2006) found that Hg concentrations in fish were correlated with Hg in wet deposition across 25 states. Atkeson et al. (2005) found that reduced Hg concentrations in largemouth bass and egret nestlings in the Everglades region of Florida were correlated with reductions in point source emissions of RGM. In Sweden regional gradients in atmospheric deposition were linked to concentration gradients in biota (Munthe et al., 2004). Other studies however, have not found any clear relationship (Evers and Clair, 2005). Recent work by (Sunderland et al., 2009) in the eastern North Pacific Ocean found total Hg concentrations were higher than previously measured and suggested that oceanic circulation, and increasing atmospheric deposition due to sources in Asia, are linked to this increase. They also suggested this increase could impact Hg concentrations in pelagic marine fish.

SOURCE ATTRIBUTION FOR Hg
DEPOSITED TO THE UNITED STATES

Estimating the source contributions of Hg deposited to the United States requires the use of regional and global models. Global models are best applied to assess impacts of long-range transport because boundary conditions for regional models will affect results (Bullock and Jaeglé, 2008), although one approach that allows the use of more detailed regional models is to apply global models to set the boundary conditions (Seigneur et al., 2004; Bullock and Jaeglé, 2008). As with all models the accuracy of the underlying meteorological data will affect results. Uncertainties that influence Hg modeling results include inadequate understanding of Hg atmospheric chemistry, magnitude of sources and sinks, limited data on speciation of anthropogenic emissions and on dry deposition, and the significance of Hg recycling between the air and terrestrial and marine surfaces (Seigneur et al., 2004; Lin et al., 2006; Jaeglé et al., 2008; Pirrone, 2008).

To understand the impact of long-range transport of Hg on deposition to the United States we first need to understand the impacts associated with domestic sources. Figure 4.2 shows the percentage of domestic anthropogenic source contribution to U.S. deposition, based on the GEOS-CHEM model (Selin et al., 2007). This figure shows that in the eastern United

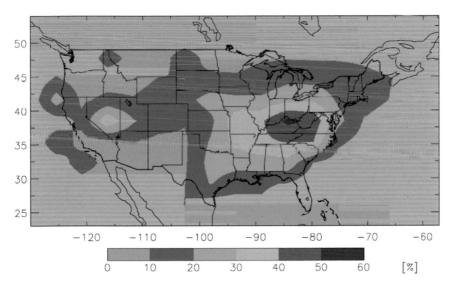

FIGURE 4.2 Percent contribution of anthropogenic North American emissions to total (wet + dry) deposition in the United States
SOURCE: Selin et al., 2007.

States, where most anthropogenic point sources of reactive and particle-bound Hg are located, domestic sources may contribute up to 60 percent to total deposition. For the western United States, the estimated contribution of domestic anthropogenic sources to deposition is much smaller. For both regions the remaining deposition (based on model results) is derived from regional and global natural emissions, current foreign anthropogenic emissions, and legacy Hg in the active pool.

Other models give similar results, although the details of these models (i.e., emission estimates, atmospheric reactions, dry deposition rates) vary. For example, Seigneur et al. (2004) applied a global chemical transport model (CTM) and a nested continental CTM, and found that across North America, depending on location, 10 to 80 percent of deposition was due to domestic anthropogenic emissions, with an area average of 25-32 percent; and Asian anthropogenic emissions contributed approximately 20 percent. The spatial variation for the latter was significant (5 to 36 percent), being higher in the west and declining across the continent. In this model, the deposition associated with natural sources also exhibited a large range (6 to 59 percent), with this source being more important in the western United States.

Travnikov (2005) used a hemispheric model (MSCE-Hg-Hem) to assess the long-range transport and deposition of Hg to North America. Results indicated that Asia contributed 24 percent, North America contributed 33 percent, and Europe contributed 14 percent (percentages include both natural and anthropogenic contributions). In this assessment North America was found to contribute 5 percent to the deposition in Europe and 4 percent to the deposition in Asia. More recently, Strode et al. (2008) applied the GEOS-Chem model and found that Asian and North American sources each contributed ~ 25 percent to deposition in the United States (including all sources), or 14 and 16 percent, respectively, if only current anthropogenic sources were considered.

Selin et al. (2008) applied the GEOS-Chem model in more detail and found that 20 percent percent of North American deposition was generated internally from anthropogenic sources, while 30 percent was derived from external anthropogenic sources. The remaining 30 percent was from natural sources and 20 percent from legacy Hg. Model simulations explain 50 to 65 percent of the variance of the MDN wet deposition data over the United States (Seigneur et al., 2004; Bullock and Jaeglé, 2008; Selin et al., 2008).

As part of the recent UNEP initiative (Pirrone and Mason, 2009a), four global models were used to assess the impact of a 20 percent anthropogenic emission reduction in four source regions: South Asia, East Asia, Europe, and North America. In all regions, the greatest effects of the emissions reductions occurred within the region itself, largely because of the significant contribution of RGM and Hg_p was associated with point sources.

Reducing East Asian anthropogenic emissions by 20 percent resulted in a ~ 3 percent reduction of deposition to North America. In the longer term, a 20 percent reduction would have a greater impact because all emissions add to the global pool of Hg that is continually recycled among environmental reservoirs.

KEY FINDINGS AND RECOMMENDATIONS

Question: What is the role of long-range transport on Hg deposi tion to the United States?

Finding. Once emitted from any source, Hg has the potential to be transformed to different chemical forms, transported through the atmosphere, and deposited long distances from the point of origin. Hence, long-range transport is an important process that clearly affects U.S. exposures. Continued emissions will increase the amount of Hg in the global pool available for long-range transport and recycling between reservoirs.

Finding. Mercury deposition to the contiguous United States has increased since the beginning of the industrial revolution, as anthropogenic releases have increased. Recent modeling studies suggest a range of 10 to 80 percent of the Hg deposited to the United States is from domestic anthropogenic sources (depending on location) with an average of ~ 30 percent for the country as a whole. The rest is derived from natural sources, foreign anthropogenic emissions, and the active legacy pool.

Finding. Key limitations for understanding long-range Hg transport and its exchange among different reservoirs include our knowledge of atmospheric chemical processing, dry deposition, and the potential for Hg deposited by wet and dry processes to be emitted back to the atmosphere. Evaluation of model results is limited by the availability of key observational data.

Recommendation. Pursue further research focused on understanding Hg atmospheric chemistry and kinetics, the dry deposition of Hg species, the global spatial (horizontal and vertical) and temporal variability in air concentrations and deposition, and potential for recycling of current and legacy Hg. Develop methods that allow for the measurement of concentrations and deposition of Hg that are easy to apply and do not require elaborate field stations. We also recommend efforts to improve emission inventories for natural and anthropogenic sources, including the magnitude and speciation of emissions in sectors where there is significant uncertainty. All such studies will be useful for improving and verifying model results. Model sensitivity analyses would be useful for further defining research needs.

Question: What are the potential implications of deposition of Hg from the global pool with respect to human health, ecosystems, and air quality management goals?

Finding. Hg exposure for humans occurs primarily through fish consumption and is a legitimate health concern for certain human populations (i.e., pregnant women and small children who consume fish and those that consume large quantities of fish with high Hg concentrations). Methyl Hg exposure may also adversely impact wildlife. A component of Hg in fish is derived from the global atmospheric Hg pool and therefore long-range transport is an important contributor to Hg found in U.S. ecosystems

Finding. Hg will continue to be deposited to the United States in the future, with the magnitude of these inputs depending on foreign and domestic emission controls. Point source controls can reduce local deposition of Hg, especially the more reactive forms; however, because Hg is globally distributed and recycled between surfaces and the atmosphere after being released from a source, major reduction in Hg deposition cannot occur without cooperation on a global scale.

Recommendation. Because the endpoint of concern for human exposure to Hg is fish consumption, research that focuses on development of a better understanding of the linkage between atmospheric Hg deposition and the means by which Hg enters aquatic food webs is needed.

Question: What factors might influence future Hg emissions and the potential for long-range transport?

Finding. It is expected that coal will continue to be an important and growing source of energy for much of the world, especially in developing countries. The implications for Hg emission and deposition will depend upon the rate of emissions growth, the specific type of coal used, and the pollution control technologies employed. Hg emissions from other sectors (such as mining, incineration, and industrial operations) may also increase, as economic growth leads to greater extraction and processing of natural resources. Pollution control technologies that remove Hg (II) or Hg_p may have some local benefit; however, without control of Hg (0) emissions, the globally available pool will continue to increase.

Finding. Climate change could affect the global cycling of Hg in a number of ways, for instance

• warming temperatures could increase the release of Hg to the air from soils and oceans;
• increased frequency of forest fires would increase the emissions of legacy Hg;
• enhanced plant growth (resulting from elevated atmospheric CO_2 levels) could increase Hg uptake from the air;
• more rapid plant decomposition may result in release of the Hg stored in leaf litter;
• changes in atmospheric circulation patterns could affect the dynamics of Hg transport and deposition;
• changes in atmospheric oxidant concentrations (related to climate change or other causes) could affect the patterns of Hg deposition.

Predicting the impact of climate change on the global Hg cycle is challenging given the uncertainties associated with impacts on climate, weather, and atmospheric dynamics, and the complexity of air-surface-plant Hg exchange.

Recommendation. The United States should actively engage in international cooperation for reducing Hg emissions worldwide, including efforts to advance and globally disseminate technologies that reduce Hg emissions associated with energy use and other forms of industrial activity.

Recommendation. Continue research designed to understand the Hg biogeochemical cycle and potential implications of climate change on the global Hg cycle. This is especially important given the large pool of Hg in terrestrial and oceanic compartments and the possibility of increased mobilization of this Hg.

5

Persistent Organic Pollutants

SOURCES AND GLOBAL REGULATORY STATUS

In the Stockholm Convention on Persistent Organic Pollutants (POPs), a chemical is considered persistent (from an atmospheric standpoint) if it has been measured at locations distant from sources of potential concern, if monitoring data show that long-range atmospheric transport may have occurred, or if modeling results show that the chemical has a potential for long-range atmospheric transport, with an atmospheric half-life exceeding two days (http://www.pops.int/documents/convtext/convtext_en.pdf). Under typical windspeeds, a chemical can travel 150-800 km in two days and result in contamination in remote locations (Scheringer, 2009).

Estimated atmospheric gas-phase reaction half-lives of POPs (Table 5.1) range from less than one day to greater than a year. In general, POPs with gas-phase reaction half-lives greater than two days and POPs absorbed onto fine particulate matter are capable of undergoing long-range atmospheric transport. The reaction half-lives of POPs in other environmental media (including water, soil, and sediment), are considerably longer than in the atmosphere. Because of their long reaction half-lives in water, soil, and sediment, and their potential for revolatilization and atmospheric transport, some POPs cycle in the global environment for many decades, similar to mercury.

In this chapter, we focus on the POPs identified in the United Nations Economic Commission of Europe's Convention on Long-range Transboundary Air Pollution (UNECE LRTAP) (http://www.unece.org/env/lrtap/pops_h1.htm), which include aldrin, chlordanes, chlordecone, DDT,

TABLE 5.1 Some Persistent Organic Pollutants (based on reaction with hydroxyl radical)

POP	Use/Source	Year Banned in U.S.	Atmospheric Gas-Phase Half-Life[a] (Days)
Aldrin	Insecticide	1987	0.17
Chlordane	Insecticide	1988	2.1
Chlordecone	Insecticide	1978	> 365
DDT	Insecticide	1972	3.1
Dieldrin	Insecticide	1987	1.2
Heptachlor	Insecticide	1988	0.2
Hexabromobiphenyl	Flame retardant	1973	38
HCB	Fungicide	1984	> 365
Mirex	Insecticide	1978	> 7
PCBs	Industrial	1977	3 to 120
Toxaphene	Insecticide	1980	4.7
HCHs (technical mixture)	Insecticide	1978	19
PAHs	Combustion	NA	0.5 to 2
PCDD/Fs	Combustion	NA	2.6 to 200

[a]EPA EpiSuite: http://www.epa.gov/oppt/exposure/pubs/episuitedl.htm.

dieldrin, heptachlor, hexabromobiphenyl, hexachlorobenzene (HCB), mirex, polychlorinated biphenyls (PCBs), toxaphene, hexachlorocyclohexanes (HCHs), polycyclic aromatic hydrocarbons (PAHs), and polychlorinated dibenzo-p-dioxins and furans (PCDD/Fs) (Table 5.1) (http://www.unece. org/env/lrtap/pops_h1.htm). This list includes products of incomplete combustion (PAHs and PCDD/Fs), pesticides (aldrin, chlordanes, chlordecone, DDT, dieldrin, heptachlor, hexachlorobenzene, mirex, toxaphene, and HCHs), and industrial chemicals (PCBs and hexabromobiphenyl). It is noted that while the use of the synthetic organic POPs such as pesticides and industrial chemicals was banned in the United States several decades ago, they continue to volatilize from historically-contaminated soils and cycle in the environment. In addition, the emission of combustion derived POPs (PAHs and PCDD/Fs) may be reduced through combustion emission control but cannot be realistically eliminated.

The objective of the UNECE POPs protocol is to eliminate any emission, discharges, or losses of POPs. The protocol bans the use and production of some POPs (aldrin, chlordane, chlordecone, dieldrin, endrin, hexabromobiphenyl, mirex, and toxaphene), has severe restrictions on the use of some POPs (DDT, HCHs, and PCBs), and plans for a scheduled elimination of other POPs (DDT, heptachlor, HCB, PCBs) at a later date. In addition, the protocol requires parties to reduce emissions of HCB,

PCDD/Fs, and PAHs below 1990 emissions. The United States was a signatory party to the protocol in 1998 but has not ratified it (http://www.unece.org/env/lrtap/status/98pop_st.htm).

Although persistent organic pesticides are banned or restricted in most developed countries throughout the world, they continue to cycle in the global environment due to revolatilization from historically contaminated soils, vegetation, and water bodies. As a result, some persistent organic pesticides continue to undergo long-range atmospheric transport, deposition, and bioaccumulation in remote U.S. high-elevation and high-latitude ecosystems (Clarkson, 2002; Kucklick et al., 2002, 2006; Howe et al., 2004; Vander Pol et al., 2004; Kannan et al., 2005; Hageman et al., 2006; Muir et al., 2006; Su et al., 2006, 2008; Usenko et al., 2007; Ackerman et al., 2008; Landers, 2008). Over long periods of time (years to decades) these compounds eventually degrade or become sequestered in deep soils and sediments. As a result, atmospheric concentrations of persistent organic pesticides are generally decreasing in remote U.S. ecosystems (Sun et al., 2006b, 2007; Usenko et al., 2007; Landers, 2008).

In contrast, the atmospheric concentrations of POPs emitted during incomplete combustion (including particulate-phase PAHs and PCDD/Fs) have generally remained the same or increased in remote locations, due to increased global combustion (Prevedouros et al., 2004; Becker et al., 2006; Sun et al., 2006c; Usenko et al., 2007; Venier et al., 2009). In addition, there has been a significant shift in the emission of PAHs (and possibly PCDD/F emission) from developed countries to developing countries (Zhu et al., 2008; Zhang and Tao, 2009). As of 2004, Chinese and Indian emissions of 16 PAHs were significantly greater than U.S. emissions (114 Ggy^{-1}, 90 Ggy^{-1}, and 32 Ggy^{-1} for China, India, and the United States, respectively) (Zhang and Tao, 2009). This suggests that given the expected increased energy demands in developing countries, human and ecosystem exposure to PAHs (and possibly PCDD/Fs) may increase in coming years, especially in the rapidly developing regions of the world, while exposure to the other identified POPs may stay the same or decrease.

TOXICOLOGICAL RELEVANCE

POPs typically have low water solubility, high lipid solubility, and an intrinsic resistance to natural degradation processes. Because of these properties, POPs are environmentally persistent and tend to bioaccumulate in adipose tissue, putting breast-feeding infants at higher risk of adverse health effects. The individual compounds have different toxicological properties that can result in adverse biological effects in fish, wildlife, and humans. A wide range of adverse health outcomes has been associated with exposure to individual POPs (see Agency for Toxic Substances and Disease Registry reviews of chlordane, DDT, aldrin/dieldrin, HCH, heptachlor, mirex, PAHs, PPBs. PCBs, and toxaphene). Potential human health effects include impairment of the immune system, nervous system, hormonal system, and reproductive functions. Concerns over the risks of POPs have led to the establishment of worldwide monitoring programs to determine concentrations of POPs in adipose tissue and associated adverse consequences (Jorgenson, 2001; Li et al., 2006b).

A Joint WHO and Convention Task Force on the Health Aspects of Air Pollution assessed the health risks of priority POPs in relation to long-range transboundary pollution (WHO, 2003). The risks associated with the following groups of substances were reviewed: pentachlorophenol, DDT, hexachlorocyclohexanes, hexachlorobenzene, heptachlor, polychlorinated dibenzo-p-dioxins and dibenzofurans, polychlorinated biphenyls, and polycyclic aromatic hydrocarbons. The task force also performed a short hazard assessment for polychlorinated terphenyls, polybrominated diphenylethers, polybrominated dibenzo-p-dioxins and dibenzofurans, and short-chain chlorinated paraffins to identify the main gaps in information necessary for risk assessment.

ATMOSPHERIC FATE AND INTERMEDIA TRANSPORT

The more volatile POPs (such as HCB and HCHs) exist primarily as gases in the atmosphere, while less volatile POPs exist primarily in atmospheric particulate phases (including most PCDD/Fs). Still other POPs (such as PAHs) are distributed between both the atmospheric gas and aerosol phases. As a result, POPs are subject to both gas-phase and aerosol-phase removal mechanisms in the atmosphere, including wet and dry deposition, gas exchange, and direct and indirect photolysis. Precipitation, particularly snow, is an efficient scavenger of POPs from the atmosphere (Wania et al., 1999; Halsall, 2004; Lei and Wania, 2004). When this snow melts, it releases a pulse of POPs to the surrounding ecosystem (Daly and Wania, 2004; Lafreniere et al., 2006; Meyer et al., 2006).

Gas-phase POPs react with photochemically generated OH radical in the atmosphere, and has been shown to be the most significant environmental transformation reaction for some POPs (Anderson and Hites, 1996; Mandalakis et al., 2003). In general, the photochemical degradation rate decreases (and atmospheric half-life increases) for aerosol-phase POPs, increasing the potential for long-range atmospheric transport of POPs sorbed to fine particles. The reaction of PAHs with NO_x and O_3 are significant in that these reactions may result in the formation of more toxic nitro and oxy-PAHs (Pitts et al., 1985; Helmig et al., 1992; Sasaki et al., 1997).

Because POPs are emitted primarily from anthropogenic combustion, industrial, and agricultural sources, their atmospheric concentrations tend to decrease with distance from populated areas in the United States (Hafner et al., 2005). The concentrations of some POPs have been shown to increase in remote areas due to episodic transport events from source regions (Hageman et al., 2006; Usenko et al., 2007; Primbs et al., 2008a,b; Genualdi et al., 2009). The global fractionation of some POPs to high latitudes (Simonich and Hites, 1995) and the "cold trapping" of some POPs at high elevations has also been observed (Blais et al., 1998; Landers et al., 2008).

EMISSION INVENTORIES

The development of global emission inventories for POPs is challenging, especially for synthetic organic POPs, because of the lack of information on how much of a chemical was manufactured and used throughout the world. Historically, many POPs have been deposited or applied to soils and continue be released as secondary emissions (similar to mercury). These secondary emissions are difficult to estimate, given that the magnitude and distribution of the original deposition and resulting emissions are largely unknown. Thus, global emission inventories for POPs have high degrees of uncertainty. Global emission estimates have been developed only for PCBs (Breivik et al., 2002a,b, 2007), HCHs (Li et al., 2003; Su et al., 2006), HCB (Barber et al., 2005) and PAHs (Zhang and Tao, 2009). European emission estimates exist for DDT, HCB, PAHs, and PCDD/Fs (Pacyna et al., 2003a). In many cases POP emissions from developing countries are largely unknown.

U.S. INFLOW AND OUTFLOW OF POPS

In recent years the inflow of POPs from Eurasia to the western United States via transpacific atmospheric transport has been identified and documented (Killin et al., 2004; Primbs et al., 2008a,b; Genualdi et al., 2009). Although strong transpacific transport events are episodic, occurring primarily in the late winter and spring, the inflow of POPs from Eurasia to the western United States likely occurs at a low level throughout the year. In particular, elevated concentrations of aerosol-phase PAHs, HCB, and alpha-HCH have been measured in transpacific air masses relative to regional North American air masses at remote sites in the Pacific Northwestern United States (Killin et al., 2004; Primbs et al., 2008a,b; Zhang et al., 2008a; Genualdi et al., 2009). Figure 1.3 shows an example of the transport of several POPs to Mt.Bachelor in Oregon, along with several other indicators of Asian anthropogenic sources (Primbs et al., 2008a,b). These same POPs have also been measured in outflow from Asia (Primbs et al., 2007). Other studies show outflow of PAHs from China (Guo et al., 2006; Lang et al., 2008). Studies have documented the emission of PAHs and re-emission of synthetic organic POPs from soils and vegetation during the large 2003 Siberian fires and the subsequent transpacific transport of these emissions to two sites in the western United States (Genualdi et al., 2009). In a recent study conducted to identify POP source regions and emissions from recent use vs. re-emissions from historic use, one chemical form of HCH (racemic alpha-HCH) was measured in Asian, transpacific, and free tropospheric air masses, while a different chemical form (nonracemic alpha-HCH) was measured in regional U.S air masses (Genualdi, 2009). Nonracemic alpha-HCH is indicative of re-emissions from historic use (and "aged" signature) because of enantioselective biodegradation by soil organisms over time.

Our current understanding of the magnitude of the inflow of POPs to the United States through transpacific transport of Eurasian emissions is limited but growing. Hageman et al. (2006) estimated the relative contribution of regional (within 150 km radius) and long-range (> 150 km radius) atmospheric transport to the dieldrin, alpha-HCH, chlordane, and HCB concentrations in annual snow pack collected from remote, high-elevation sites in seven western U.S. national parks, by correlating their measured concentrations with the cropland intensity within 150 km of the park. They estimated that 100 percent of the POP concentrations measured in Alaskan parks were due to long-range transport, while 30 to 70 percent of the concentrations of these POPs measured in the most westerly continental U.S. park (Mt. Rainier National Park) were due to long-range transport (including transpacific transport). At progressively more interior parks (Glacier and Rocky Mountain National Parks) the contribution from long-range transport decreased to 10 to 30 percent.

In comparison, the inflow of POPs from Canada to the United States, Mexico to the United States, and West Africa to the southeastern United States is much less documented. Gamma-HCH (lindane) used in the Canadian prairie provinces (Saskatchewan, Alberta, and Manitoba) has been shown to be transported to the Great Lakes (Ma et al., 2003, 2004) and Glacier National Park (Hageman et al., 2006). In 2004 elevated POP concentrations were measured at a remote mountain site in the western United States and linked to emissions from forest fires in western Canada (Primbs et al., 2008a,b). Although high concentrations of POPs have been measured in air throughout Mexico (Wong et al., 2009), their direct atmospheric transport from Mexico to the United States has not been documented. Saharan dust storms have the potential to transport POPs from Africa to the southeastern United States (Zhang et al., 2008b; Pozo et al., 2009), but this phenomenon has not been well characterized either.

The outflow of POPs from the United States to the Great Lakes region, and the resulting bioaccumulation in this ecosystem, has long been recognized and modeled (Hafner and Hites, 2003, 2005; Ma et al., 2005c,d).

The Integrated Atmospheric Deposition Network (IADN), a joint US and Canada monitoring program, has been in place since the 1990s (Glassmeyer et al., 1997, 2000; Hillery et al., 1998; Hites, 1999; Cortes et al., 2000; Buehler et al., 2002, 2004; James and Hites, 2002; Hafner and Hites, 2003, 2005; Carlson et al., 2004; Sun et al., 2006a,c, 2007; Venier et al., 2009). Atmospheric clearance rates of POPs (reported as half-lives) have been calculated from the long-term monitoring record for gas-phase PAHs, PCBs, HCHs, chlordanes, dieldrin, HCB and DDT, and range from 1 to 32 years (Sun et al., 2006b,c, 2007). These data suggest that the atmospheric concentrations of these POPs are decreasing in the Great Lakes region. Aerosol-bound PAH and PCDD/F concentrations have not decreased significantly over the same period at these remote sites (Sun et al., 2006c; Venier et al., 2009), perhaps as result of increased combustion worldwide (Zhang and Tao, 2009).

Numerous studies have documented the outflow of POPs from the United States to the Canadian Arctic, and since 1992, atmospheric measurements of POPs have been made at Alert, a monitoring station in the Canadian Northwest Territories (Hung et al., 2001, 2002a,b; Prevedouros et al., 2004; Becker et al., 2006; Su et al., 2006, 2008). In comparison, there is much more limited data on atmospheric POP concentrations and source regions to the U.S. Arctic (only Point Barrow, Alaska, from 2000 to 2003) (Su et al., 2006, 2008). Because of the low temporal resolution of the Point Barrow dataset (samples collected over the period of a week), a thorough investigation into the geographic location of POP source regions to the U.S. Arctic has not been conducted. Many studies have, however, documented the deposition of POPs in Alaska (Hageman et al., 2006), and their bioaccumulation in food webs (Kucklick et al., 2002, 2006; Howe et al., 2004; Vander Pol et al., 2004; Kannan et al., 2005; Ackerman et al., 2008).

Although the outflow of POPs from the United States to Mexico and Europe likely occurs to some degree, this outflow has not been directly measured in discrete air masses and its magnitude is unknown. Atmospheric POP concentrations have been measured in remote Western Europe, including Mace Head, Ireland (Lee et al., 1999, 2004), but air masses with elevated POP concentrations were primarily tracked back to other parts of Europe.

EXISTING POP MODELING CAPABILITIES

Because of their global atmospheric transport potential, distribution between the atmospheric gas and aerosol phases, and potential to partition to and from various environmental media, global atmospheric transport models for POPs are not as refined as the global transport models for PM species or volatile compounds. Modeling the long-range transport of POPs requires a global emission inventory, a transport model, and a detailed understanding of removal processes. Because each POP has unique chemical properties and sources, global modeling has been conducted for a limited number of POPs.

In North America most POP modeling has focused on regional sources to the Great Lakes. For example, findings from Ma et al. (2005a,b) indicate episodic transport of toxaphene from the southeastern United States to the Great Lakes. Zhang et al. (2008b) showed episodic transpacific transport of lindane (γ-HCH) and transatantic transport to the Caribbean and southeastern United States from Africa, although there is little data to validate these modeling results. A connection between the concentration anomalies of several POPs in the Great Lakes with sea-urface temperature anomalies in the tropical Pacific (an indicator of El Niño) was shown by Ma and Li (2006). While this long-distance correlation is somewhat surprising at first glance, it is plausible that there is a mechanism linking El Niño with transport patterns and re-emission of previously deposited toxic compounds, such as POPs or Hg.

Gusev et al. (2007) describe global modeling activities at the Meteorological Synthesizing Centre (MSC)-East for α-HCH and PCB-153. In their analysis European, North American, and Asian emissions of PCB-153 are 65, 14, and 8 percent of the global total, respectively. Using a 20 percent reduction scenario (similar to the HTAP O_3 analyses discussed in Chapter 2), the authors show that intercontinental influence for PCB-153 is modest. A 20 percent reduction in PCB-153 emissions from Europe results in a 7, 3.5, and 2 percent decline in European, Arctic, and Asian deposition, respectively. The much smaller North American emissions of PCB-153 show minimal influence outside this region. For α-HCH, emissions are dominated by certain countries, including Algeria, Tunisia, Spain, France, Romania, North Korea, Vietnam, Maylasia, and India. European, North American, and Asian emissions of α-HCH are 26, 0.3, and 35 percent of the global total, respectively. Intercontinental transport of α-HCH is more significant than PCB-153 because it is present in the atmospheric gas phase rather than the aerosol phase. For example, a 20 percent reduction in α-HCH emissions from East Asia results in a 3 percent reduction in deposition in North America.

While these modeling exercises are an important step forward, to date there has been relatively little interaction between the modeling and observational communities. As a result, these global model predictions are largely unverified. These results demonstrate that long-range transport of POPs is an important process to consider, but we are not yet able to make accurate quantitative predictions for most compounds.

It is important to be able to model the episodic nature of long-range transport of POPs. CTMs with episodic transport prediction capability (such as GEOS-CHEM), have not yet been parameterized for POPs. Satellite observations of POPs are not currently possible because of their extremely low concentrations in the atmosphere (pg/m^3 to ng/m^3). Existing satellite images for particulate matter and gas-phase combustion products may not be appropriate surrogates for pesticides because of differences in source regions, atmospheric chemistry, and deposition.

NEW POPS

Some currently used chemicals, including pesticides (such as endosulfan) and consumer product chemicals (such as fluorinated organic chemicals [FOCs] and polybrominated diphenyl ethers [PBDEs]) are now found in remote locations throughout the world and have the potential to be considered POPs. Human exposure to the chemicals used in consumer products includes both direct exposure (skin and inhalation) and exposure via the food web; concentrations of some of these potentially new POPs have increased in the food web, human blood serum, and mother's milk in recent years in the United States (Schecter et al., 2006, 2007; Schecter, 2008; Tao et al., 2008).

Some FOCs, such as fluorotelomer alcohols (FTOHs), have been shown to undergo microbial degradation (Dinglasan et al., 2004) and photochemical transformation (Wallington et al., 2006) to perfluorinated carboxylic acids (PFCAs), including perfluorooctanoic acid (PFOA). These compounds can undergo atmospheric long-range transport (Shoeib et al., 2006; Piekarz et al., 2007) and oceanic transport (Armitage et al., 2006; Wania, 2007). Their precursors can also be found throughout the globe, including in remote locations such as Alaska and the Arctic and high-elevation ecosystems (Smithwick et al., 2006; Stock et al., 2007; Loewen et al., 2008; Schenker et al., 2008).

PBDEs are also distributed globally, including remote U.S. high-elevation and high-latitude ecosystems (Kannan et al., 2005; Muir et al., 2006; Usenko et al., 2007; Ackerman et al., 2008). Although there have been regulations and voluntary efforts to move from the persistent, bioaccumulative, and toxic tetra- and penta-brominated PBDEs to safer alternatives, these alternatives have been shown to undergo photodegradation (Hua et al., 2003; Bezares-Cruz et al., 2004; Zeng et al., 2008) and microbial degradation in the environment (He et al., 2006; Robrock et al., 2008), resulting in the same tetra- and pent-brominated PBDEs.

CLIMATE CHANGE AND POPS

Because the global environmental fate of POPs (including their degradation and intermedia transport) is highly temperature dependent, global climate change has the potential to significantly change the current global distribution of POPs. For example, increased surface temperatures could result in the volatilization of POPs from current temperate and tropical source regions and their deposition in colder regions, such as high-elevation and high-latitude ecosystems (Simonich and Hites, 1995; Wania and Mackay, 1995; Blais et al., 1998). On the other hand, if temperature increases more at higher latitudes compared with lower latitudes, fewer POPs will be stored at high latitudes.

The melting of glaciers may result in the release of POPs stored decades ago into global circulation (Donald et al., 1999; Geisz et al., 2008) and POPs currently stored in soil and vegetation, may be re-released during fire events (Primbs et al., 2008a,b; Genualdi et al., 2009). Decreases in sea ice cover and increases in ocean temperature also have the potential to result in the redistribution of the more volatile POPs stored in ocean water (Macdonald et al., 2003, 2005; Macdonald, 2005). Although increased surface temperatures would theoretically increase the degradation rate of POPs in the environment, this benefit may be offset if POPs are redistributed to colder environments with more limited sunlight.

Because rain and snow are efficient scavengers of airborne POPs, changes in precipitation patterns can affect where and how efficiently POPs are removed from the atmosphere. Because of their affinity for terrestrial surfaces, the global distribution of POPs will change along with vegetation patterns. Changes in the North Atlantic Oscillation, the El Niño-Southern Oscillation, the Arctic Oscillation, and the Pacific North American pattern could all affect the re-release and global redisribution of POPs currently stored in ocean water (Macdonald et al., 2003, 2005; Macdonald, 2005; Ma and Li, 2006). The tremendous uncertainty in predicting how POPs will be distributed globally in the future adds to the motivation for maintaining long-term atmospheric monitoring programs.

KEY FINDINGS AND RECOMMENDATIONS

Question: What do we know about the current import and export of POPs?

Finding. There is substantial observational evidence that POPs can be transported over intracontinental scales, but only a few transport pathways have been documented. For instance, transpacific atmospheric transport of POPs to the contiguous United States is relatively well characterized, whereas inflow to Alaska is not. There is evidence of inflow from Canada to the United States, while inflow from Mexico is not well characterized. Outflow of POPs from the United States to Canada and the Arctic is fairly well characterized, whereas outflow to Europe is not. At present it is not possible to quantify these transport fluxes due to the limited emission inventories and lack of validated quantitative models for POPs.

Finding. There is evidence that atmospheric concentrations of banned or restricted-use pesticide-related POPs are declining due to global regulations, while concentrations of combustion-related POPs (PAHs and possibly PCDD/Fs) are increasing due to growing emissions from developing countries. Some chemicals currently in use that have the potential to be considered POPs due to their persistence, bioaccumulation potential, and toxicity (such as PBDEs and FOCs) are known to undergo long-range transport and have exhibited increasing concentrations in the food web and humans in recent years.

Question: What are the potential implications of long-range transport of POPs on humans and ecosystems and environmental management goals?

Finding. U.S. efforts to reduce exposures to POPs are clearly impacted by long-range transport. It is difficult to characterize the significance of this influence, both because of the scientific uncertainties described above and because there are currently no clear national goals for POPs deposition.

Question: How might the factors influencing these issues change in the future?

Finding. There is potential for the U.S. population (as well as some remote high-elevation and high-latitude U.S. ecosystems) to be exposed to increasing concentrations of certain POPs that have increasing emissions outside the United States (for instance, inhalation exposure to carcinogenic PAHs and food web exposure to bioaccumulated PCDD/Fs). This potentially increasing exposure may be more pronounced in the western United States because of the patterns of transpacific transport from Asian countries.

Finding. There is potential for the enhanced re-release of legacy POPs from melting glaciers, forest fires, and warming soils and oceans due to climate change influences. There is likewise potential for remote high-elevation and high-latitude U.S. ecosystems to be exposed to increasing POP concentrations through re-release and redeposition of these compounds. The impacts of these future changes cannot currently be predicted quantitatively.

Recommendation. Improve our ability to quantify the impacts of long-range transport of POPs on human and ecosystem health and to predict how this might change in the future. This requires efforts to

- develop hemispheric emission inventories and projections for POPs (especially PAHs and PCDD/Fs).
- parameterize and validate hemispheric transport models for POPs. Future research on global transport of POPs should focus on bringing together modeling and observation specialists during the design and execution of field campaigns to identify the most important parameters to measure and therefore reduce model uncertainties.
- better quantify the current U.S. inflow and outflow of POPs through measurements and modeling. This includes the continuation of long-term atmospheric monitoring programs, which can aid our ability to track how POPs are redistributed due to climatic and global emission changes.
- estimate future U.S. inflow and outflow based on projected changes in source regions and global climate change, using hemispheric transport models.
- expand our understanding of the photochemical processes that affect POPs during transport.
- evaluate long-range transport potential in the initial assessment and regulatory approval of new chemicals.

6

Crosscutting Issues and Synthesis

The preceding four chapters each focused on a priority pollutant class and identified a wide array of requirements and opportunities for advancing our understanding of long-range transport of pollution. Many of the issues discussed are unique to particular pollutants, but there are also a number of issues that are clearly important to a wide range of pollutants. This chapter highlights a series of crosscutting issues that the committee identified for focused discussion. We then present the committee's vision of an integrated observational and modeling system to advance our understanding of long-range transport of pollution (Section 6.2), followed by an overview of key conclusions from this study (Section 6.3). This chapter is intended to build upon the findings and recommendations presented in preceding chapters, with the goal of maximizing the effectiveness of future efforts by addressing multiple pollutant classes simultaneously.

CROSSCUTTING ISSUES

New Pollutant Fingerprinting Techniques   The high degree of complexity involved in identifying and quantifying long-range pollution transport, particularly in identifying and characterizing contributions from individual pollution sources, will require new developments in analytical and observational techniques. The difficulty in resolving sources is a limiting factor in determining the contributions of long-range transport for almost all pollutants. It is particularly difficult in the case of PM since their heterogeneous composition and their origins in both primary emission and secondary formation mean that chemical transformations need to be understood as well. In the past decade there have been major advances in PM measurement techniques, but further advances will be needed to better characterize long-range transport.

Isotopic Signature Studies   High-precision, stable isotope ratio measurements are increasingly being applied to both gaseous species and aerosol PM elemental source identification. With this technique the primary identification factor resides at the submolecular level. This technique can be highly sensitive because isotopic species are reduced in abundance and therefore small variations can be detected. Using traditional analytical methods, one often cannot precisely measure the primary species of interest when its background concentration is large (e.g., carbon dioxide). Isotopic measurements do not face such limitations, and are increasingly used to study atmospheric sources and sinks of compounds such as methane, nitrous oxide, carbon dioxide, carbon monoxide, ozone, and aerosol sulfate and nitrate.Recent simultaneous development of methods to measure stable isotopes at a high precision in micro- to nanogram-size samples, and the development of new quantum mechanical theories for the mechanisms that alter stable isotope ratios, have greatly advanced isotopic techniques. A review of the application of stable isotopes for atmospheric composition studies is available (Thiemens, 2006). Precise characterizations of isotopic compositions provide a new fingerprinting technique. This is a particularly powerful application for long-range transport studies because it is one of few measurements that provides information about the chemical processes that have occurred during transit as well as their bulk properties at arrival.

For example, applications of isotopic techniques provide a means by which ship emissions can be recognized in regions with multiple pollution sources that cannot be separated by traditional concentration measurements (Dominguez et al., 2008). Long-range transport of aerosols from Asia to the United States may also be recognized from their characteristic isotopic signature, and their secondary transformations during transit may be resolved (Patris et al., 2007). Isotope ratio measurements can also play a significant role in advancing our understanding of the biogeochemistry of mercury. As discussed in Chapter 4, many important aspects of mercury chemistry are presently unresolved, such as source identification, transformation, and storage in various environmental reservoirs. Isotope measurements can be an effective way to gain new insights on these issues, if used in combination with traditional concentration and chemical speciation measurements. The measurement of mercury isotopes is challenging, and further analytical developments are needed to fully exploit this technique. In the case of POPs there have thus far been few studies at the isotope level. Future developments of carbon and deuterium isotope ratio measurements may be useful but limited, given the nonisotope specificity of sources and transformational processes.

Given recent advances in understanding the fundamental physical chemistry of isotope effects and the development of new isotope measurements (e.g., multi-isotope measurements of individual molecules; positional-specific isotopic measurements; and new measurement techniques, including in situ, real-time laser-based isotopic measurements and satellite remote sensing isotope measurements), the use of stable isotope ratio measurements could play an increasingly significant role in all aspects of understanding international pollutant transport.

Molecular Chirality Measurements A recently proposed technique involves the study of enantiomeric signatures of chiral organochlorine pesticides (OCPs). Chiral OCPs are manufactured in racemic form,[1] but selective biodegradation processes can result in preferential loss of one enantiomer over the other and a shift to a nonracemic signature (i.e., unequal amounts of left- and right-handed molecules). In these cases a racemic signature may indicate recent use of the pesticide, while a nonracemic signature may indicate past use, with selective biodegradation of certain enantiomers and revolatilization from soils. For example, in a study by Genualdi et al. (2009) racemic alpha-hexachlorocyclohexane (HCH) was measured in Asian, transpacific, and free tropospheric air masses, while nonracemic alpha-HCH was measured in regional U.S air masses, thus illustrating the potential value of such techniques for distinguishing between transpacific and regional U.S. air masses.

Single-Particle Measurements A major advancement in aerosol science has been the development and continued enhancement of the ability to do real-time measurements of single aerosol particle composition and size. Conventional techniques typically require collection and subsequent analysis of the particles, which may alter the physical and chemical characteristics. With techniques such as aerosol-time-of-flight mass spectrometry (ATOFMS) the chemical composition and size of particles can be determined in real time and without collection. Such measurements are invaluable in chemically fingerprinting aerosols and their sources with a high degree of specificity. For episodic transport episodes the capability for fast timescale resolution allows for a more detailed modeling of transport in general and chemistry in particular. New developments in this area will be increasingly important for addressing the issue of secondary surface chemical reactions.

---

[1] This refers to molecules that have equal amounts of left- and right-handed enantiomers (i.e., molecules that are reverse mirror images of each other).

Online mass spectrometry methods can also be used for the accurate apportionment of aerosols by comparing the fingerprints of individual atmospheric particles to known source signatures. Such methodologies are particularly important in source identification and quantification. They have been used to study long range transport over India and determine the major source of the Atmospheric Brown Cloud (Spencer et al., 2008), and to identify the impacts from many other sources including incineration in Mexico City and China (Moffet et al., 2008; Zhang et al., 2009b), wildfires (Mühle et al., 2007), gasoline and diesel emissions, and ship emissions (Toner et al., 2008; Ault et al., 2009). These instruments are also beginning to be used to study the climatic impacts of specific pollution sources, including impacts on the optical properties (Moffet et al., 2008) and cloud-forming potential of atmospheric particles (Cubison et al., 2008; Furutani et al., 2008). Such studies are critical for determining what pollution sources are having the largest impact on our climate, so they can be properly regulated.

New identification and characterization developments are also key in the general area of organics, especially particulate matter (Canagaratna et al., 2007), but also volatile and semivolatile species. Organic compounds may constitute between 10 and 75 percent of atmospheric aerosols by mass. Though these particles play an important role in radiative forcing and human health issues, understanding of their sources and atmospheric chemistry is poorly established. Recent advances in FTIR and X-ray scanning transmission spectroscopy have significantly improved our capacity to study such issues, especially with identifying functional groups and morphology of single particles, which may be source dependent. Two dimensional mapping of aerosol organic functional groups is emerging as a valuable technique for studying sources and transport of organics. The use of nondestructive soft X-ray beams (NEXAFS-STM) is another important emerging technique.

All these various single particle measurement techniques are important for the study of long-range transport of particulate matter. They help improve our understanding of aerosol chemistry, formation and growth processes, and subsequent impacts on atmospheric lifetime and transport properties.

Other Methods   An array of other innovative pollution tracing or fingerprinting methods are currently being developed and tested. For instance, gas and aerosol ratios have also been used to give information on emissions and chemical processing (e.g., Parrish et al., 1998). This includes methods to quantify Asian emissions of Hg (0) (Jaffe et al., 2005a; Weiss-Penzias et al., 2007), emissions of particulate mercury (Finley et al., 2009), and ozone processing during transport (Parrish et al., 1998; Price et al., 2004).

Finding. Informed decision making about long-range pollution transport requires sophisticated abilities to identify and quantify specific pollution sources. Past studies have suffered from analytical limitations and an inability to make sufficiently frequent and precise measurements. Breakthroughs in analytical fingerprinting techniques (e.g., isotopic or enantiomeric signature measurements, online measurement of single particles, aerosol or gas ratios) expand our ability to define sources and their strengths, and to allow high-resolution characterization of transport.

Recommendation. We recommend continued investment in advancing these cutting-edge fingerprinting techniques and the widespread deployment of such techniques in both ground- and air-based studies, aimed at refining our assessment of pollution sources, transport, and chemical transformation. Particular emphasis should be placed on using such techniques to advance understanding of the complex reactions of organic species (in both the gas and particle phase) and the complexities of the mercury biogeochemical cycle.

Emission Sources, Inventories, and Projections   Modeled estimates of long-range transport depend on the magnitude of present and future emissions in upwind regions. Emission fields are a limiting factor in the fidelity of these models. Some general emission inventory needs include the following:

Spatial resolution   Presently most global inventories are available at $1° \times 1°$ resolution. This has been sufficient for most global modeling studies, as it matches or exceeds the spatial resolution of most models. However, this resolution is insufficient for determining the initial evolution of continental plumes, especially for highly reactive pollutants. Both nested emission inventories and nested models, with higher resolution in high-concentration areas, are needed to more accurately represent initial plume processing and transport.

Emission projections   Understanding the effects of a changing climate and an evolving economy on future natural or anthropogenic emissions requires mechanistic representations of the factors that affect those emissions. A good match between observations and models provides confidence that an inventory is reasonably accurate, but the underlying mechanisms and driving factors must also be well represented.

Emission Inventory Evaluation   Inventory compilations have been produced to support modeling activities such as AEROCOM (Dentener et al., 2006). The utility of these collections lies in providing a common format, common inputs for model intercomparion studies, and some comparison tools. Still missing and needed are quantitative assessments of inventory quality. First, the basic assumptions in different inventories need examination. This endeavor needs to include comparison and validation of the underlying driving factors: activity levels, technology mix, and emission rates. Second, work is needed to ensure that an inventory-model combination reproduces atmospheric observations reliably. Such a match is sought in intensive measurement campaigns (Huebert et al., 2003; Warneke et al., 2007), and diagnosing the causes of mismatches is the goal of inverse modeling techniques (Mendoza-Dominguez and Russell, 2000; Kasibhatla et al., 2002). The use of satellite data for selected tropospheric pollutants, including CO and NO_2 and PM, can also be used to identify emission hot spots and evaluate regional emission inventories. Sophisticated inverse modeling, most recently utilizing adjoint methods with advanced CTMs, show great promise as top-down checks on bottom-up emission inventories (Kopacz et al., 2009; Kurokawa et al., 2009). Methods are also needed for rapidly updating emission inventories to incorporate the findings of these studies as soon as they become available.

Emission Inventory Diversity Atmospheric models have benefited from comparisons, and ensembles of such models are typically more robust than individual model results. The same diversity is frequently lacking in emission models. Despite the complexity of the responsible processes that produce emissions, single estimates of combustion, wildfire, dust, and biogenic emissions are typically used by all chemical and transport models. When separate estimates exist, they can identify discrepancies (Ma and van Aardenne, 2004). An increase in the number of contributors to emission estimates and projections may result in greater confidence in these critical inputs. At the same time, measurements need to be designed to test the causes of discrepancies.

Each emission source affects multiple pollutants and needs to be represented consistently to produce holistic estimates of changes in air quality. As discussed in Chapter 1, we separate emission sources into three categories: (i) those related to energy and economy, (ii) those associated with the natural environment, and (iii) legacy sources, or environmental reservoirs storing pollutants that can be re-emitted. There is some debate about these divisions; for example, dust lofted from disturbed soils is neither completely natural nor wholly attributable to economic activity; but we chose the classifications listed here because they correspond to broad differences between procedures in both estimating and mitigating emissions.

(i) **Energy and economy** Energy use and other economic activity include the sectors of electricity generation, manufacturing, consumer products, household energy, transportation, and agriculture. In turn, each sector may contain many source types that have very different emission rates. For example, household energy needs may be met with natural gas, which produces few emissions except NO_x, or with unprocessed solid fuels, which have high emissions of particulate matter, carbon monoxide, and volatile organic compounds. Relative contributions and technological makeup of each sector vary between regions. HTAP-TF (2007) provides a comprehensive summary of emission inventories that represent energy and economy. Rather than reprising that review here, we describe general methods of producing those inventories and identify directions required to advance them further.

Three general approaches, or a combination thereof, can be used to develop regional or global emission fields. The first method requires data from individual facilities, which are either measured by continuous emission monitors or self-reported. These data have the greatest potential for accuracy, although challenges certainly exist in both monitoring and reporting. Such data are available only from large point sources in regions with strong regulations, including North America, Europe, Japan, and increasingly, China.

In a second approach the emission estimate is based on the technological makeup within each emitting sector, including abatement technologies. This approach is used by many global and regional inventories, including the RAINS (Regional Air Pollution Information and Simulation) model based at International Institute for Applied Systems Analysis (IIASA) (Kupiainen and Klimont, 2007), inventories for TRACE-P (Streets et al., 2003) and individual countries (Reddy and Venkataraman, 2002), and some global models of individual pollutants (Bond et al., 2004). The supporting technology descriptions may be obtained from records, from expert judgment, or by extrapolation from other regions. These estimates benefit greatly from involvement of country experts.

The last method requires the least detail: applying a single emission factor to an entire emitting sector. These emission factors may depend on region, with more polluting activities assumed in regions with less development or lower technology. This was the approach used in EDGAR 2.0 (Emissions Database for Global Atmospheric Research) (van Aardenne et al., 2001) and many other global inventories. EDGAR is currently being updated to a technology-based approach.

Because future emissions entail both activity changes and technology changes, treatment of single sectors provides only a qualitative sense of future emissions. Nevertheless, it is the method presently used by many integrated assessment models (van Vuuren et al., 2006). Studies consider technology change in a simple fashion by applying principles of the "Environmental Kuznets Curve," in which cleaner activity is assumed to follow development. However, some studies have questioned the support for this simple relationship (Deacon and Norman, 2006), as development may occur more quickly than expected based on history (Stern, 2004). Thus, hybrid models that link technological change and economic growth are desirable.

Detailed emission projections in which technological change is linked with economic conditions are well under way for large point sources and for on-road transportation. Such projections are available in regions with experience in environmental regulations, such as the United States, Europe, and China. Modeling is less advanced for sources that are more dispersed, including off-road vehicles, agricultural emissions, waste burning, and residential combustion.

In all cases current inventories and future projections are less developed and more uncertain when a history of regulation is lacking. International working groups such as those convened under NARSTO and the United Nations Environment Program (UNEP) "Atmospheric Brown Cloud" are needed to share knowledge about driving factors and to build capacity for inventory estimation in developing countries. Emissions of air pollutants differ from those of greenhouse gases, particularly CO_2, because they may be highly dependent on the combustion or manufacturing process. The sectoral divisions used for reporting inventories for the United Nations Framework Convention on Climate Change (UNFCCC) tend to be inadequate for air quality purposes because they do not identify these individual processes.

The evolving trajectories of global and regional economies are one of the greatest uncertainties in projecting air pollutant emissions. Inherent uncertainties are likely to persist even with additional understanding of the factors that cause and mitigate emissions. Modeling and regulatory approaches need to acknowledge and account for such future uncertainties.

(ii) **Natural environment** Contributions from the natural environment are observed in both episodic transport and in regional and transported background concentrations. These emissions originate from dust storms, forest fires, vegetation, soil, freshwater, and ocean surfaces, all of which depend greatly on meteorology, land use, and climate. Because of the latter two factors, anthropogenic activity plays a role in these emissions.

Estimating emissions from this assortment of natural sources requires representation of biogeochemical connections between the emission source and the environment. Both atmospheric variables and human interactions govern the emitting capacity—for example, whether barren soil has the potential to emit dust, the amount of biomass in grassland that can be burned, or the prevalence of broadleaf trees that may emit terpenes. Meteorology also affects whether and when the emission occurs—for example, dust lofting with increased wind speed, increased terpene emissions at higher temperatures, the correlation of Hg soil emissions with temperature, solar radiation, and soil moisture. Such fundamental relationships are used to predict both present-day and future emissions.

Challenges in modeling emissions include large seasonal variability and the need to simulate sources in remote environments with little available data. Onsets of episodic events, such as fire ignition, are nearly impossible to predict. Thus, models of present-day emissions combine process modeling with inputs from remote sensing (Reid et al., 2004; Guenther et al., 2006; van der Werf et al., 2006; Kalashnikova and Kahn, 2008). Examples of satellite data incorporated include burned area and new fire and dust plume events.

Progress in projecting future emissions from the natural environment will require challenging and confirming the fundamental relationships in emission models. This evaluation should occur both at large scales, using remote sensing data and intensive field experiments, and at microphysical scales, for example, with flux chamber, eddy correlation, and relaxed eddy accumulation measurements. Statistical characterization that covers magnitudes and frequencies of emission events is needed to assess future impacts on both background and episodic exceedances. Unusual events such as volcano eruptions also affect atmospheric chemistry, and no effort is made to predict these.

(iii) **Legacy emissions** Legacy emissions result when pollutants are re-emitted from environmental reservoirs. Predicting these sources is the most complex of all, as it requires modeling not only the re-emission process but also the history of inputs to the reservoir and their removal or re-emission over time. Despite the importance of these emissions for toxic and environmentally stable pollutants, modeling of legacy emissions is still in its infancy (Bergan et al., 1999; Lohmann et al., 2007).

Finding. Consistency in modeling long-range transport requires a source-based, multi-pollutant treatment of emission inventories. These inventories are fundamental to the fidelity of models that estimate present-day and future long-range transport. High spatial resolution is needed to represent reactive plumes at their origins.

Finding. Improvements in present-day emission inventories will be possible only by representing the factors that modulate those emissions. These factors are as diverse as the response of the natural environment to changing climate and shifts in the mix of technology caused by tightened emission standards.

Finding. Understanding of global cycling for very long-lived pollutants, such as mercury or POPs, is limited in part by the lack of emission inventories that estimate re-emission from environmental reservoirs (legacy emissions).

Finding. Traditional bottom-up emission inventories need to be periodically validated by top-down inverse CTM modeling of measured atmospheric concentration fields. Airborne multipollutant concentration measurements are ideal when available; satellite measurements are also becoming important and offer data for some key pollutants that are frequently updated, allowing assessment of temporal variations in emissions. The use of adjoint methods with state-of-the-science CTMs shows great promise for improving the accuracy of inverse model results.

Recommendation. Enhance the ability to understand, forecast, and manage changing emission sources by designing field experiments that not only confirm emission totals but also link them to the fundamental sources and processes that generate them. These connections will decrease uncertainty in projections of emissions from both economic activity and the natural environment.

Recommendation. Promote national and international efforts to improve Northern Hemisphere regional and national emission inventories' accuracy, timeliness, spatial and temporal resolution, multipollutant coverage, and intercomparability. Also stimulate the collection, evaluation, and use of airborne and satellite multipollutant concentration data and its analysis using state-of-the-science inverse CTM modeling techniques, to independently evaluate the accuracy, completeness, and temporal and spatial fidelity of available emission inventories.

Recommendation. Enhance the understanding of legacy emissions of Hg and POPs by modeling environmental flows and reservoirs, in conjunction with historical emission inventories that cover several decades of human activity.

Meteorological Processes   Chemical transport models as well as trajectory calculations and particle dispersion models rely on meteorological conditions to simulate pollutant transport. Meteorological processes lift pollutants from the planetary boundary layer into the free troposphere, can transport them for great distances, and determine where and when they will sink back into the boundary layer to influence local air quality. If meteorological conditions are not properly represented in a chemical transport model, no amount of sophisticated chemistry can overcome the errors in pollutant transport that will result. Therefore, future advances in understanding and quantifying long-range transport of pollution will be closely linked to advances in meteorology and will require the collaborative efforts of meteorologists and atmospheric chemists.

Additional data are needed to accurately describe global meteorological conditions at even the synoptic scale, as well as the mesoscale phenomena that are embedded therein. Relatively little data are available over the oceans and less developed countries. As a result circulation centers, wind patterns, and transport pathways such as warm conveyor belts can be incorrectly located; these errors compound in time as trajectories and other tracer parameters are calculated. Meteorological observations are used for diagnostic studies and to initialize meteorological models. Initializing models with accurate meteorological data is a fundamentally important step in making numerical forecasts. Further modeling accuracy can be achieved, particularly in retrospective simulations, if four-dimensional data assimilation (FDDA) is used. FDDA is a process by which gridded or point observations are ingested into a running simulation to nudge the simulated variables toward the observed values.

Additional vertical profiles of temperature, humidity, and winds are needed over large areas of the globe. Although various satellite sensors and retrieval algorithms have been developed to provide this information, their major limitation is crude vertical resolution and often their dependence on first-guess information. Further advances in satellite sensor technology and algorithm development are sorely needed to help fill the existing data gaps. Unmanned aerial vehicles are another potential source of this needed information.

In addition to needing meteorological data, the models meant for chemistry transport studies need to be run at high resolution in an attempt to incorporate the myriad of circulations (see Appendix B) that affect long-range pollutant transport. For example, to represent the effects of topography, horizontal resolutions of < 10 km may be necessary. Variations in topography can block pollutant transport, lift the pollution to higher altitudes, or channel it through valleys. Mesoscale thermally induced mountain and valley circulations can greatly influence transport by modifying the depth of the PBL. Thus, pollutants released in areas of complex terrain may be transported quite differently than if released over flat terrain. High-resolution topographic data are available, but the meteorological models need sufficient resolution to fully utilize the data. Accurately representing these processes on an intercontinental scale will require either horizontal grid resolutions of 10 km or less (especially near the release areas) through the use of embedded high-resolution grids within coarser domains [nested grids] or through the use of parameterization schemes.

The PBL is the source of most pollutants, and even at high resolution (1-10 km) the complex, turbulent meteorological processes occurring within it (even over flat terrain) cannot be explicitly resolved by numerical models; they need to be parameterized based on larger-scale meteorological variables. Although current procedures produce reasonable results in many cases, models of pollutants sometimes represent release into the free troposphere inaccurately in both space and time. Even if the sign of this process is correctly parameterized, its magnitude may not be.

Finding. Major improvements in current abilities to model several aspects of pollution transport are needed, including

1. A more complete understanding of PBL physics to allow the development of parameterization schemes that are sufficiently robust to perform in a range of locations. Although many PBL physics schemes exist, many are based on midlatitude atmospheric dynamics and may not be suitable in either polar regions or the tropics. The goal of a PBL scheme in a numerical model is to represent the transfers of momentum, moisture, and heat in the PBL as well as the free atmosphere. Thus, inadequate PBL schemes can quickly degrade meteorological and air quality forecasts. Since these schemes rely heavily on a land surface model (LSM) to represent heat and moisture transfer from the land surface, the LSMs also affect turbulent mixing within the PBL. Therefore, we have to understand land surface dynamics around the world to aid the PBL schemes. LSMs require high-resolution land use and vegetation data from satellites in order to be effective. Advances in this area will improve PBL and LSM procedures, thereby preventing the degradation of model performance.

2. Models need to better represent the processes occurring at horizontal resolutions of 5 to 10 km. Convective transport, whether dry or associated with thunderstorms, is a prime example that directly affects pollutant transport. Many convective parameterization schemes have been designed for grid resolutions greater than 10 km. Since none of these is uniformly best in every situation, it often is difficult to decide which is best in a given situation. At grid resolutions less than 10 km, parameterization schemes either may aid or hinder air quality forecasts depending on their treatment of the convective vertical motions and the subsequent vertical redistribution of atmospheric pollutants. Very high-resolution cloud-resolving models only can be used only over very limited areas because of their computational expense. It is vital that meteorological models be able to properly depict convection

since it is a major source of pollutant transport out of the PBL and into the free troposphere where stronger winds are located.

3. Considerable pollution is released in urban settings that contain street canyon flows, radiational heating due to buildings, and obstacle effects that can temporarily trap pollutants within the PBL. In addition, nonlocal pollutants passing over urban landscapes can encounter enhanced mixing due to the rough surface. These effects may require grid resolutions of tens of meters in order to avoid the need for parameterization. At a typical mesoscale grid length (i.e., 10-20 km), the urban landscape is not explicitly resolved, requiring sophisticated schemes to account for the subgrid-scale processes. Such schemes currently are under development, but few have been implemented in operational forecast models.

Recommendation. Develop a better understanding of the basic submesoscale dynamic processes involved in the entrainment or detrainment, long-range transport, and deposition of pollutants. Utilize focused field studies, advances in satellite technology and improved data assimilation methods to continue to enhance meteorological and transport modeling capabilities. To be most effective meteorologists and atmospheric chemists, including both modelers and measurement specialists, should collaborate on efforts to obtain the phenomenological insights and data required to develop the better measurement techniques and improved numerical models that will enable us to adequately quantify the role of distant sources on local air quality.

International Pollutant Transport from Maritime Shipping and Aviation Sources Pollutant emissions from international ship and air transport are of direct interest to those studying long-range atmospheric pollutant transport because they represent a unique class of international pollutant transport, which is growing rapidly and may in some cases mimic long-range atmospheric transport. At a minimum, receptor sites (especially those near the coast) attempting to identify and characterize intercontinental atmospheric transport need to be able to distinguish between ship and aircraft emissions and emissions from distant sources.

Ship Emissions It is well recognized that marine vessels that consume sulfur-rich bunker fuels are significant sources of carbonaceous (soot) PM and gaseous and particulate sulfur (Dominguez et al., 2008; Dalsøren et al., 2009; Lack et al., 2009). The regulation of these sources at present is not stringent, although significant efforts are under way to strengthen international regulatory controls. The particulate emissions from ships using high sulfur fuels are predominantly submicron-size sulfate and carbonaceous materials; recent measurements in the Atlantic and Gulf of Mexico reported average PM concentrations of 46 percent sulfate, 39 percent organics, and 15 percent black carbon (Lack et al., 2009). Ship PM emissions directly influence Earth's albedo by reflection and absorption of sunlight, as well as by nucleating ship-track clouds; but their effect on global climate is thought to be small, relative to other aerosol sources. The most significant aspect of maritime emissions is the impact upon human health. Recent estimates suggest that as many as 60,000 deaths per year worldwide may be linked to ship emissions (Corbett et al., 2007). The economic cost of these health impacts to the United States alone may surpass $500 million annually (Gallagher and Taylor, 2003).

Ship PM emissions are capable of long-range transport and, coupled with the international aspects of maritime commerce and the predicted increase in global maritime activity, the consequences for health and economics of global coastal community is significant. As discussed in Dominguez et al. (2008) there are significant challenges associated with the resolution and quantification of local vs. long-range transported aerosols from ships. The issue is unique because it involves nonpoint sources and is subject to international regulations. As a consequence, better understanding of the nature, composition, and transport of ship emissions is highly desirable.

In addition to PM, ships emit significant quantities of CO_2, SO_2, CO, NO_x, and VOCs (Dalsøren et al., 2009). Sulfur dioxide, nitrogen oxides, and volatile organic compounds all contribute to secondary aerosol PM formation; the latter two are also tropospheric ozone precursors. While ship emissions are not a major source of mercury or manufactured POPs, ship-emitted soot and gaseous emissions are a significant source of PAH, an important class of POPs.

Recent measurements of particulate and gaseous emissions from large diesel ships have further characterized ship emissions. It is directly observed that three morphological species of varying chemical composition are emitted. These include soot, which may be metal infused, mineral dust, sulfate, and a variety of carbonaceous particulate matter (Moldanová et al., 2009). This recent work also illustrated the extraordinary chemical complexity of ship emissions. Of particular importance is the observed linkage between PAH and PM. The PM composition was largely organic carbon, with trace amounts of elemental carbon, but analyses also revealed the presence of a complex mixture of PAHs. This work demonstrates that not only are ship emissions significant sources of both PM and PAHs but also may be distinct from other transportation emissions.

Aircraft Emissions   As discussed in IPCC reports global aviation transport involves direct emissions of fine particles into the upper troposphere and lower stratosphere (IPCC, 1999, 2007). In addition, volatile PM and ozone precursors, including SO_2, CO, NO_x, and VOCs, are emitted during both takeoff, landing, and cruise modes (IPCC, 1999; Royal Society, 2008). Both PM and selected gaseous aircraft emissions near airports are subject to regulation by individual nations, using standards set by the International Civil Aviation Organization. Emissions during cruise phases of aircraft however currently are not subject to regulation.

In general, aerosol number densities are elevated in areas of active aircraft commerce, and the major impact is their influence on the formation of persistent contrails and associated cirrus clouds. The mechanism and magnitude of the associated effects, however, are not well resolved. The consequences of the ground-level addition of sulfate and soot particles from jet fuel combustion appears to be modest in comparison to local background levels (IPCC, 2007). Ice nucleation in the upper troposphere and lower stratosphere, with subsequent transport and chemical reaction on ice surfaces is thought to be one of the largest concerns, although there are no estimates of the direct impact on cloud ice particle formation rates at present.

> **Finding.** Emissions from ocean shipping and passenger and cargo aircraft are a source of long range pollution transport that can complicate the detection and characterization of long-range atmospheric pollutant transport from historical land-based sources.

> **Recommendation.** Studies of long-range atmospheric transport of pollution should be coordinated with studies of ship and aircraft cruise emissions, with the goal of determining methods to distinguish among these pollutant sources in source attribution studies.

STRENGTHENING INTEGRATED SYSTEMS FOR TRACKING AND ATTRIBUTION OF LONG-RANGE TRANSPORT OF POLLUTION

From an air quality management perspective there is a fundamental need to measure the ambient levels of pollutants discussed in this report and how their values change over time. There is a further need to relate pollution concentrations and their trends to emission sources and their changes. This information is needed, for example, to assess the effectiveness of current emission control strategies such as fuel desulfurization to reduce ambient levels of SO_2 and acid deposition fluxes, and NO_x emission reductions under the Clean Air Interstate Rule (CAIR) to reduce O_3 and PM.

To assess the extent to which national emission control strategies will be effective in meeting environmental targets, related to species of interest in this report (i.e., O_3, PM, Hg, and POPs), it is necessary to partition the concentration observed at a specific location into the part arising from national emissions (both local and regional), the part attributable to international sources, and the part coming from natural sources. As this analysis is carried out for more distant sources attribution becomes more difficult, since their contributions typically are weaker and more difficult to distinguish from local or regional components. As discussed throughout this report both observation- and model-based approaches exist for source attribution. These techniques have been developed and applied largely for source attribution applications at local and, more recently, regional scales. These techniques are now being extended for applications at the hemispheric and global scales. The model-based approaches at these scales suffer from large computational uncertainties (e.g., those related to global emissions and transport processes over long distances). Larger-scale observation-based approaches are limited due a variety of factors, including the loss of source signal over long distances, small numbers of relevant and representative observation sites devoted to source attribution, and the lack of a comprehensive suite of source detection and characterization measurements at available sites.

Improving our capability to assess the impacts of long-range transport of pollution on the effectiveness of national control strategies requires better quantification of the global distribution of pollutants and their trends, and an increased ability to perform source attribution analyses at hemispheric and global scales. The importance and required attributes of a global air quality observational strategy were articulated in an earlier NRC report *Global Air Quality: An Imperative for Long-Term Observational Strategies* (NRC, 2001), which found that current observational systems were not adequate for characterizing many important medium- and long-term global air quality changes. That report made the following recommendations.

• Maintain and strengthen the existing measurement programs that are essential for detecting and understanding global air quality changes. High priority should be given to programs that aid in assessing long-term trends of background ozone and PM.
• Establish new capabilities to provide long-term measurements and vertical profiles of reactive compounds and PM that will allow meaningful examination of long-range transport and trends in background concentrations.

These two recommendations are just as valid today as they were when issued in 2001. Some important surface-based trend measurements for ozone, PM and their precursors have continued; with some additional capacity developed (e.g., with the U.S. EPA's NCORE network [http://www.

epa.gov/ttn/amtic/ncore/], as discussed in NRC [2008]). Satellite-borne instruments, especially from NASA's Earth Observing System (EOS), have become active and produce multiyear tropospheric vertical column density measurements for some key pollutants. U.S. federal funding for new or improved pollutant trend and transport measurements was sparse during the period between 2001 and 2009, so little new capability has been deployed, or even planned, over that time frame. There is now concern that the current satellite pollutant observation capability is declining as many of the EOS instruments reach and pass their nominal life spans with no replacements in sight.

Improving our capabilities to quantify the contributions from specific source regions and sectors over the geographical regions of concern (e.g., EPA-designated air management regions, the Arctic) requires capabilities beyond those identified above for characterizing global air quality. It requires a strategy that both improves the individual components needed to perform source attributions and more closely integrates these components. As stated throughout this report there remain large uncertainties in the calculation of ambient pollution distributions and source-receptor relationships. Emissions, transport, chemical transformation, and removal processes are all major sources of uncertainty, and the relative contribution of these processes to overall uncertainty varies by pollutant. Reducing the uncertainty in predictions requires a better understanding of these processes. In addition, more critical tests of the ability of models to reproduce the processes governing intercontinental transport and the source-receptor relationships are needed. Reducing these uncertainties requires combining model and measurement techniques in an integrated analysis of data from an enhanced observing system.

Figure 6.1 depicts the major components of the needed system, including: emission inventories; chemical transport modeling; long term ground based monitoring; satellites; and intensive field studies. The specific capabilities and needs for improving the various components have been articulated earlier in this report. Here we focus on the overall system and in particular the benefits to be derived from closer integration of the different parts of this system, for instance,

- closer integration of satellite observations with emissions and modeling components (through data assimilation and inverse modeling) will lead to better emission estimates in terms of spatial distribution and absolute magnitude;
- closer integration of satellite- and ground-based observations will reduce uncertainties in observation-based spatial distributions;
- closer integration of long-term, ground-based observations with emissions elements will produce a better understanding of geographical

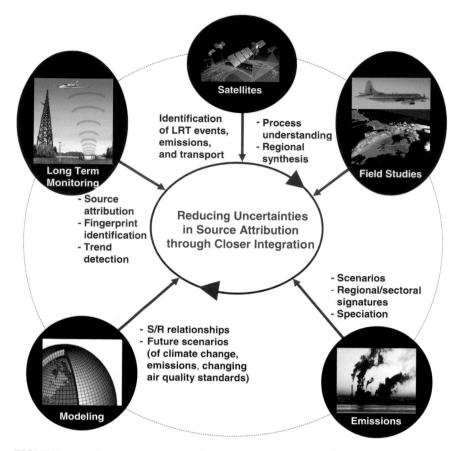

FIGURE 6.1 Major components of an integrated approach for source attribution. The circles denote the major components, the accompanying text indicates the major source and attribution-related outputs. Model and measurement elements refer to both meteorological and chemical components.

source signals, which can be exploited in fingerprint and observation-based source attribution analyses;

• closer integration of ground-based observations and models will lead to reduced uncertainties as a result of more critical evaluation of model predictions, better calibration and evaluation of observation- and model-based attributions, and improved measurement strategies;

• closer integration of field experiments with elements focused on source attribution will lead to improved understanding of processes affecting the source-receptor relationships, better techniques to follow targeted air masses (e.g., from specific source sectors or regions) over long distances;closer integration of all these components will reduce the uncertainties in current estimates of the contribution of international sources to local air quality and more accurate assessments of the impacts of imported pollution on human health and agriculture.

It is important to realize that global source attribution has not been a major design consideration in current monitoring strategies. There are additional activities needed for source attribution beyond those needed for compliance and detection of long-term trends in ambient concentrations. These additional needs can largely be met by complementing existing activities. For example, the recent report *Observing Weather and Climate from the Ground Up: A Nationwide Network of Networks* (NRC, 2009) describes a national observational infrastructure that will improve capabilities to observe and predict meteorology and air quality at the national level. It calls for enhancing the vertical dimension of key meteorological and chemical observations, increasing the number of observation sites, and the inclusion of additional chemical parameters that will help in the integration of satellite and surface observations. This plan calls for improved measurement technologies in current observational platforms (such as ground-based air quality monitoring networks, commercial aircraft, and balloons or sondes), in an enhanced lidar network, in new instrument platforms such as supersites for measuring a comprehensive suite of chemical species in remote locations, and in unmanned aerial vehicles for long-duration sampling of the atmosphere over a wide range of altitudes. This network would provide a wealth of information needed for chemical data assimilation and improved prediction of transport and chemical processes, which are needed to reduce uncertainties in source attribution estimates.

For source attribution applications the above envisioned mesoscale network would need to be complemented by additional observations at designated source attribution sites. At these sites the additional measurements would include, for instance, all of the key pollutants for which source attribution is deemed important (e.g., O_3, PM, Hg, and POPs), and a suite of observations with source attribution applications, including gas-phase PAN, speciated NMHCs, and other trace gases, selected isotopic ratios, and aerosol parameters, including composition as a function of size and single particle elemental analysis. This breadth of observations is needed to capture source signals over timescales from days to weeks and from specific source sectors (e.g., biomass burning, coal combustion). The source attribution network could be built upon exisiting long-term monitoring sites (such as Mauna Loa), with additional sites created in regions where the assessment of global contribution is needed, both within and outside the contiguous United States (e.g., Alaska). This would include elevated sites that can routinely sample free tropospheric air and are relatively free from local influences. A few of these observation sites could be identified as source attribution supersites, where more detailed and exploratory observations are taken, and where there is an embedded research program dedicated to the integration of these observations with modeling, emissions, satellite, and field study components, all focused on source attribution applications. Further details on measurement parameters and locations need to be determined by an expert group charged with designing the source attribution network.

In a similar manner improvements in our satellite observing capabilities are also needed. Currently satellite observations can provide long-term global records of a limited number of key atmospheric trace gas and aerosol parameters; observations of long-range and intercontinental transport of pollution from continental source regions; column or partial column retrievals with very limited sensitivity to the lowermost troposphere; and top-down emissions estimates on fairly coarse scales. To improve their use in long-range transport studies the following are needed:

- improvements in resolution
  —spatial coverage to facilitate the tracking of individual plume events.
  —spatial resolution in source and receptor regions to isolate city-scale features.
  —temporal resolution to capture diurnal variation of photochemistry and emissions especially in source and receptor regions. (The difficulty achieving this from polar orbit provides motivation for geostationary satellite placement).
  —vertical profile information to isolate surface sources; characterize PBL venting and entrainment of pollution; and track the transport of thin plumes. This requires multispectral and multiangle retrieval techniques and improved coverage from surface- and space-based lidar.
- improved information on aerosol size distribution and chemical speciation.
- improved retrievals over cold and bright surfaces.

Research into the closer integration of observations and models has increased over the last few years (Carmichael et al., 2008b). A greater number of models are engaged in advanced chemical data assimilation, and some are moving toward operational status (e.g., the European Union's Global and Regional Earth-system Monitoring using Satellite and in situ data [GEMS project, http://gems.ecmwf.int/]). Further efforts are needed to evaluate and refine the application of these assimilation approaches for source attribution. These approaches also need to be applied in the design of new observing systems, including new remote sensing platforms such as geostationary satellites. To accomplish these goals, model and emission inventory components need to be refined and aligned with observational strategies.

Ultimately a global network of source attribution observation sites is needed. A reasonable starting point for reaching this long-term goal is to better coordinate existing networks (see Appendix C). Programs such as the World Meteorological Organization's Global Atmosphere Watch (WMO/GAW) have made substantial efforts to establish a global network of observational sites including sites in undersampled, remote regions around the world, supported by data centers and quality control programs to enhance integration of air quality measurements from different national and regional networks.

The source attribution analysis described here requires substantial international cooperation and open exchange of data from national emission inventories and air quality monitoring networks. This effort fits well within the international WMO/GAW Integrated Global Atmospheric Chemistry Observations (IGACO) component of the Integrated Global Observing Strategy (IGOS), a partnership of international organizations concerned with global environmental-change issues, which links research, long-term monitoring and operational programs (WMO, 2004).

Building bridges between experts in emissions, meteorological and chemical modeling, satellite remote sensing, in situ observing, and the different pollutant expert communities is critical for carrying out the integrated science envisioned in Figure 6.1. An important consideration in the design and execution of a source attribution analysis will be strategies to build and strengthen connections between these communities. Programs like the International Global Atmospheric Chemistry project can play important roles in coordinating international activities that address a variety of important issues related to global air quality.

Finding. Improving our capability to assess the effectiveness of national control strategies requires better quantification of the global distribution of pollutants and their trends (as changes in the global background will impact national strategies and their effectiveness) and an increased ability to perform source attribution analyses at global scales.

Finding. Improving our capabilities to quantify the contributions from specific source regions and sectors of concern requires a strategy that improves the individual components needed to perform source attributions, and that more closely integrates these components.

Finding. Global source attribution has not been a major design consideration in current monitoring strategies. Thus there are additional activities needed for source attribution beyond those for compliance and detection of long-term trends in ambient concentrations.

Finding: Source attribution analysis, which builds upon a global observing system with integration among other key elements, requires substantial international cooperation and an open exchange by countries of data from national emission inventories and air quality monitoring networks.

Finding. Building bridges among experts in emissions, meteorological and chemical modeling, satellite and in situ observing, and the different pollutant research communities is a critical requirement to carry out the integrated science envisioned.

Recommendation. An integrated source attribution program should be established to help assess the contribution of distant sources to U.S. air quality and to evaluate the effectiveness of national control strategies to meet environmental targets. The program should focus on improving capabilities (and reducing uncertainties) within the areas of emission measurements and estimates, atmospheric chemical and meteorological modeling, long-term ground-based observations, satellite remote sensing, field experiments, and impact assessments—and integrating these components as effectively as possible to focus on attribution to particular sources and regions.

Recommendation. An expert group should be established to help design this source attribution network (e.g., suggesting parameters to be measured, identifying appropriate monitoring sites, developing an embedded research program). These efforts should take into consideration the need for international cooperation and the opportunities to collaborate within existing international efforts such as the WMO/GAW International Global Atmospheric Chemistry Observation program.

SUMMARY OF KEY MESSAGES

Strong evidence from both observations and modeling studies confirms that the pollutants considered in this study are transported over long ranges, both to and from North America, and that U.S. environmental goals are affected to varying degrees by nondomestic sources of these pollutants. It is also clear that our ability to characterize such impacts is currently limited, due to uncertainties related to source strengths, chemical transformations during transport, transport mechanisms, rates of exchange between the boundary layer and free troposphere, and a number of other important issues.

Under present global socioeconomic scenarios nondomestic influences on air quality are expected to increase in the future and will likely be an issue of increasing concern. Enhancing observations, chemical transport models, trend analyses, studies of reaction mechanisms for relevant species, and emission inventories and projections will all be of critical importance to better quantify such effects.

The pollutants explored in this study do not represent all species of concern, but they do illustrate the variability of pollutant composition and behavior and provide a focused target for analyzing the phenomenon of long-range transport. To improve our quantitative understanding of such issues, we recommend that the United States strongly support more extensive international cooperation in research (observational and modeling) and assessment, and ultimately in emissions control efforts.

The four classes of pollutants analyzed in this study, the Committee found a number of crosscutting shortcomings in our capabilities for observation, analysis, emissions prediction, and transport dynamics.

Observations Ambient concentrations vary significantly for the different species of interest, but in all cases, their temporal and spatial variations are not sufficiently understood, and observations of a significantly greater resolution are needed. This includes both ground- and aircraft-based monitoring and satellite observations. Each pollutant species has significantly different characteristics, including chemical reactivity, atmospheric creation and loss processes, and phase change properties. In all cases, however, there is a critical need for measurements of sufficient temporal and spatial resolution to allow better quantification of long-range transport. There is also a consistent need for observational platforms that allow one to discern between episodic and nonepisodic events, and between remote contributions to background pollution vs. less frequent, large perturbation events. In many cases (e.g., for mercury) there are major gaps in our understanding of species' chemical fate in the atmosphere and in other environmental reservoirs. Comprehensive measurements are the only viable means to fill these gaps and provide the quantitative data required for informed decision making.

Emission Inventories There is a significant variation in the adequacy of existing inventories. None could be described as fully adequate, but some (such as mercury) are particularly poorly resolved, as a result of environmentally complex chemical interactions. Some pollutant sources, such as ships and aircraft, that are inherently international in nature require more sophisticated and extensive observations and analysis. Emission inventories may be improved in some cases by development of additional satellite observations, assuming continued development and launch of advanced sensors and suitable platforms designed to increase the vertical and horizontal resolution, increased species coverage, and higher sensitivities for important tropospheric pollutants. The geographic extent of long-range transport episodes precludes the option of relying solely on in situ (ground- or aircraft-based) observations. Long-range transport models and inventories at the continental and hemispheric scales will increasingly rely on remote sensing observations, coupled with inverse models that help to better define fluxes from specific source regions.

Modeling Capabilities   While a number of sophisticated chemical transport models exist, expanded capabilities are crucial for understanding and analysis of current observations, developing accurate projections of future changes, and supporting informed decision making regarding long-range transport of pollution. For example, observed changes in baseline ozone concentrations and some episodic ozone pollution events have evaded systematic understanding and description. Models are needed that can better quantify the processes of boundary-layer exchange and plume dispersion for all species addressed in this report. The scales of such modeling analyses need to range from local to hemispheric, if the issue of long-range transport and the associated perturbations are to be fully understood. There also needs to be significant interaction between observational networks and modeling efforts, to help guide sampling and measurement protocols and for calibration and sensitivity analysis of new models. More pervasive and robust methods for estimating the uncertainties in model results need to be developed and implemented.

The various pollutants addressed in this report differ in their consequences for human health and the environment, but in each case there is potential for significant and growing adverse impacts. We see a clear need to enhance scientific understanding of these pollutants' interactions with the environment, including their long-range transport, entrainment and deposition fluxes, and ultimate impacts on human health and ecosystems. Addressing issues of long-range pollutant transport requires a significant degree of international cooperation.

As a final note, the Committee wishes to emphasize the obvious fact that our planet shares one atmosphere. Pollution emissions within any one country affect populations, ecosystems, and climate properties well beyond national borders. Measures taken to decrease emissions in any one region can have benefits that are distributed across the Northern Hemisphere. The United States, as both a source and receptor of this long-range pollution, has a responsibility to remain actively engaged in addressing this issue.

It is clear that local pollution can be affected by global sources, although in most cases air quality violations are driven by local emissions. But regardless of where the pollution originates, protecting human and ecological systems from dangerous levels of pollution should be the policymakers' primary objective. Meeting this objective will require strengthening domestic pollution control efforts to whatever levels are required to ensure that a population's total pollution exposure (from local, regional and distant sources) does not exceed safe levels. However, reducing the impacts of distant emissions on local air quality cannot be achieved by domestic efforts alone. Cooperative international action, to advance our understanding of long-range transport of pollution and its impacts, and to use that understanding to effectively control emissions from both domestic sources—needs to be vigorously pursued. The Committee hopes that the analyses and recommendations in this report will help stimulate and guide those actions.

References

ACAP (Arctic Council Action Plan to Eliminate Pollution of the Arctic). 2005a. Arctic Mercury Releases Inventory: Reduction of Atmospheric Mercury Releases from Arctic States. Copenhagen, Denmark: Danish Ministry of the Environment, Danish Environmental Protection Agency.

———. 2005b. Assessment of Mercury Releases from the Russian Federation. Copenhagen, Denmark: Danish Ministry of the Environment, Danish Environmental Protection Agency.

Ackerman, L.K., A.R. Schwindt, S.L.M. Simonich, D.C. Koch, T.F. Blett, C.B. Schreck, M.L. Kent, and D.H. Landers. 2008. Atmospherically deposited PBDEs, pesticides, PCBs, and PAHs in western US national park fish: Concentrations and consumption guidelines. Environmental Science & Technology 42(7):2334-2341.

Ahrens, C.D. 2007. Essentials of Metoeorology: An Invitation to the Atmosphere. Florence, KY: Brooks Cole.

Al-Saadi, J., J. Szykman, R.B. Pierce, C. Kittaka, D. Neil, D.A. Chu, L. Remer, L. Gumley, E. Prins, L. Weinstock, C. MacDonald, R. Wayland, F. Dimmick, and J. Fishman. 2005. Improving national air quality forecasts with satellite aerosol observations. Bulletin of the American Meteorological Society 86(9):1249-1261.

Amyot, M., G. Southworth, S.E. Lindberg, H. Hintelmann, J.D. Lalonde, N. Ogrinc, A.J. Poulain, and K.A. Sandilands. 2004. Formation and evasion of dissolved gaseous mercury in large enclosures amended with $^{200}HgCl_2$. Atmospheric Environment 38(26):4279-4289.

Anderson, B.T., and A. Strahler. 2008. Visualizing Weather and Climate. New York: John Wiley & Sons.

Anderson, P.N., and R.A. Hites. 1996. Oh radical reactions: The major removal pathway for polychlorinated biphenyls from the atmosphere. Environmental Science & Technology 30(5):1756-1763.

Andreae, M. O., and P. Merlet. 2001. Emission of trace gases and aerosols from biomass burning. Global Biogeochemical Cycles 15(4):955-966.

Anenberg, S., J.J. West, A.M. Fiore, D.A. Jaffe, M.J. Prather, D. Bergmann, C. Cuvelier, F.J. Dentener, B.N. Duncan, M. Gauss, P. Hess, J.E. Jonson, A. Lupu, I.A. MacKenzie, E. Marmer, R.J. Park, M. Sanderson, M. Schultz, D.T. Shindell, S. Szopa, M.G. Vivanco, O. Wild, and G. Zeng. 2009. Intercontinental impacts of ozone pollution on human mortality. Environmental Science & Technology (submitted).

Angevine, W.M., M.P. Buhr, J.S. Holloway, M. Trainer, D.D. Parrish, J.I. MacPherson, G.L. Kok, R.D. Schillawski, and D.H. Bowlby. 1996. Local meteorological features affecting chemical measurements at a north Atlantic coastal site. Journal of Geophysical Research–Atmospheres 101(22):28935-28946.

Ariya, P., K. Peterson, G. Snider, and M. Amyot. 2009. Mercury transformations in the gas, aqueous, and heterogenous phases: State of the art science and uncertainties. In Mercury fate and transport in the global atmosphere emissions, measurements and models, edited by N. Pirrone and R. Mason. 644, Berlin: Springer Verlag.

Armitage, J., I.T. Cousins, R.C. Buck, K. Prevedouros, M.H. Russell, M. MacLeod, and S.H. Korzeniowski. 2006. Modeling global-scale fate and transport of perfluorooctanoate emitted from direct sources. Environmental Science & Technology 40(22):6969-6975.

Atkeson, T.D., C.D. Pollman, and D.M. Axelrad. 2005. Recent trends in hg emissions, deposition, and biota in the Florida everglades: A monitoring and modeling analysis. In Dynamics of mercury pollution on regional and global scales: Atmospheric processes and human exposures around the world, edited by N. Pirrone and K. R. Mahaffey. 637-656. New York: Springer.

Ault, A.P., M.J. Moore, H. Furutani, and K.A. Prather. 2009. Impact of emissions from the Los Angeles port region on san diego air quality during regional transport events. Environmental Science & Technology (ASAP)(0).

Auvray, M., and I. Bey. 2005. Long-range transport to Europe: Seasonal variations and implications for the European ozone budget. Journal of Geophysical Research–Atmospheres 110(11):1-22.

Bader, M.J. 1995. Images in weather forecasting: A practical guide for interpreting satellite and radar imagery. New York: Cambridge.

Bagnato, E., A. Aiuppa, F. Parello, S. Calabrese, W. D'Alessandro, T.A. Mather, A.J.S. McGonigle, D.M. Pyle, and I. Wängberg. 2007. Degassing of gaseous (elemental and reactive) and particulate mercury from mount etna volcano (southern italy). Atmospheric Environment 41(35):7377-7388.

Barber, J.L., A.J. Sweetman, D. van Wijk, and K.C. Jones. 2005. Hexachlorobenzene in the global environment: Emissions, levels, distribution, trends and processes. Science of the Total Environment 349(1-3):1-44.

Barrie, L.A. 1986. Arctic air pollution: An overview of current knowledge. Atmospheric Environment–Part A General Topics 20(4):643-663.

Bates, T.S., T.L. Anderson, T. Baynard, T. Bond, O. Boucher, G. Carmichael, A. Clarke, C. Erlick, H. Guo, L. Horowitz, S. Howell, S. Kulkarni, H. Maring, A. McComiskey, A. Middlebrook, K. Noone, C.D. O'Dowd, J. Ogren, J. Penner, P.K. Quinn, A.R. Ravishankara, D.L. Savoie, S.E. Schwartz, Y. Shinozuka, Y. Tang, R.J. Weber, and Y. Wu. 2006. Aerosol direct radiative effects over the northwest atlantic, northwest pacific, and north indian oceans: Estimates based on in situ chemical and optical measurements and chemical transport modeling. Atmospheric Chemistry and Physics 6(6):1657-1732.

Becker, S., C.J. Halsall, W. Tych, H. Hung, S. Attewell, P. Blanchard, H. Li, P. Fellin, G. Stern, B. Billeck, and S. Friesen. 2006. Resolving the long-term trends of polycyclic aromatic hydrocarbons in the Canadian arctic atmosphere. Environmental Science & Technology 40(10):3217-3222.

Bein, K.J., Y.J. Zhao, M.V. Johnston, and A.S. Wexler. 2008. Interactions between boreal wildfire and urban emissions. Journal of Geophysical Research–Atmospheres 113(D07304).

Bell, M.L., F. Dominici, and J.M. Samet. 2005. A meta-analysis of time-series studies of ozone and mortality with comparison to the National Morbidity, Mortality, and Air Pollution Study. Epidemiology 4:436-445.

Bell, M.L., R.D. Peng, and F. Dominici. 2006. The exposure–response curve for ozone and risk of mortality and the adequacy of current ozone regulations. Environmental Health Perspectives 114:532-536.

Bell, M.J., A. McDermott, S.L. Zeger, J.M. Samet, and F. Dominici. 2004. Ozone and short-term mortality in 95 US urban communities, 1987 2000. Journal of the American Medical Association 292(19):2372-2378.

Bergan, T., L. Gallardo, and H. Rodhe. 1999. Mercury in the global troposphere: A three-dimensional model study. Atmospheric Environment 33:1575-1585.

Bertschi, I.T., and D.A. Jaffe. 2005. Long-range transport of ozone, carbon monoxide, and aerosols to the ne pacific troposphere during the summer of 2003: Observations of smoke plumes from Asian boreal fires. Journal of Geophysical Research–Atmospheres 110(D05303).

Bezares-Cruz, J., C.T. Jafvert, and I. Hua. 2004. Solar photodecomposition of decabromodiphenyl ether: Products and quantum yield. Environmental Science & Technology 38(15):4149-4156.

Biester, H., R. Bindler, A. Martinez-Cortizas, and D.R. Engstrom. 2007. Modeling the past atmospheric deposition of mercury using natural archives. Environmental Science and Technology 41(14):4851-4860.

Bindler, R., I. Renberg, P.G. Appleby, N.J. Anderson, and N.L. Rose. 2001. Mercury accumulation rates and spatial patterns in lake sediments from West Greenland: A coast to ice margin transect. Environmental Science and Technology 35(9):1736-1741.

Biswas, A., J.D. Blum, B. Klaue, and G.J. Keeler. 2007. Release of mercury from rocky mountain forest fires. Global Biogeochemical Cycles 21(1).

Blais, J.M., D.W. Schindler, D.C.G. Muir, L.E. Kimpe, D.B. Donald, and B. Rosenberg. 1998. Accumulation of persistent organochlorine compounds in mountains of western Canada. Nature 395(6702):585-588.

Bloom, N.S. 1992. On the chemical form of mercury in edible fish and marine invertebrate tissue. Canadian Journal of Fisheries and Aquatic Sciences 49(5):1010 1017.

Bond, T.C., D.G. Streets, K.F. Yarber, S.M. Nelson, J.-H. Woo, and Z. Klimont. 2004. A technology-based global inventory of black and organic carbon emissions from combustion. Journal of Geophysical Research 109:D14203, doi:10.1029/2003JD003697.

Bortz, S.E., M.J. Prather, J.-P. Cammas, V. Thouret, and H. Smit. 2006. Ozone, water vapor, and temperature in the upper tropical troposphere: Variations over a decade of MOZAIC measurements. Journal of Geophysical Research 111:D05305.

Bousquet, P., P. Ciais, J.B. Miller, E.J. Dlugokencky, D.A. Hauglustaine, C. Prigent, G.R. Van der Werf, P. Peylin, E.-G. Brunke, C. Carouge, R.L. Langenfelds, J. Lathière, F. Papa, M. Ramonet, M. Schmidt, L.P. Steele, S.C. Tyler, and J. White. 2006. Contribution of anthropogenic and natural sources to atmospheric methane variability. Nature 443:439-443, doi:10.1038/nature05132.

Braune, B.M., M.L. Mallory, and H.G. Gilchrist. 2006. Elevated mercury levels in a declining population of ivory gulls in the Canadian arctic. Marine Pollution Bulletin 52(8):978-982.

Breivik, K., A. Sweetman, J.M. Pacyna, and K.C. Jones. 2002a. Towards a global historical emission inventory for selected pcb congeners—a mass balance approach 1. Global production and consumption. Science of the Total Environment 290(1-3):181-198.

———. 2002b. Towards a global historical emission inventory for selected pcb congeners—a mass balance approach 2. Emissions. Science of the Total Environment 290(1-3):199-224.

———. 2007. Towards a global historical emission inventory for selected pcb congeners—a mass balance approach-3. An update. Science of the Total Environment 377(2-3):296-307.

Browning, K.A., and G.A. Monk. 1982. A simple model for the synoptic analysis of cold fronts. Quarterly Journal, Royal Meteorological Society 108(456):435-452.

Browning, K.A., and N.M. Roberts. 1994. Structure of a frontal cyclone. Quarterly Journal–Royal Meteorological Society 120(520):1535-1557.

Brunke, E.G., C. Labuschagne, and F. Slemr. 2001. Gaseous mercury emissions from a fire in the cape peninsula, South Africa, during January 2000. Geophysical Research Letters 28(8):1483-1486.

Buehler, S.S., I. Basu, and R.A. Hites. 2002. Gas-phase polychlorinated biphenyl and hexachlorocyclohexane concentrations near the great lakes: A historical perspective. Environmental Science & Technology 36(23):5051-5056.

———. 2004. Causes of variability in pesticide and pcb concentrations in air near the great lakes. Environmental Science & Technology 38(2):414-422.

Bullock, O.R. Jr., and L. Jaeglé. 2008. Importance of a global scale approach to using regional models in the assessment of source-receptor relationships for mercury. In Mercury fate and transport in the global atmosphere measurements, models and policy implications, edited by N. Pirrone and R. Mason. 377-388. New York: Springer.

Burger, J., and M. Gochfeld. 2004. Mercury in canned tuna: White versus light and temporal variation. Environmental Research 96(3):239-249.

Butler, T., M.D. Cohen, F.M. Vermeylen, G.E. Likens, D. Schmeltz, and R.S. Artz. 2007. Regional precipitation mercury trends in the eastern USA 1998-2005 Atmospheric Environment 42:1582-1592.

Butler, T.M., M.G. Lawrence, B.R. Gurjar, J. van Aardenne, M. Schultz, and J. Lelieveld. 2008. The representation of emissions from megacities in global emission inventories. Atmospheric Environment 42(4):703-719.

Caldwell, C.A., P. Swartzendruber, and E. Prestbo. 2006. Concentration and dry deposition of mercury species in arid south central New Mexico (2001-2002). Environmental Science and Technology 40(24):7535-7540.

Canagaratna, M.R., J.T. Jayne, J.L. Jimenez, J.D. Allan, M.R. Alfarra, Q. Zhang, T.B. Onasch, F. Drewnick, H. Coe, A. Middlebrook, A. Delia, L.R. Williams, A.M. Trimborn, M.J. Northway, P.F. DeCarlo, C.E. Kolb, P. Davidovits, and D.R. Worsnop. 2007. Chemical and microphysical characterization of ambient aerosols with the aerodyne aerosol mass spectrometer. Mass Spectrometry Reviews 26(2):185-222.

Cape, J.N. 2008. Surface ozone concentrations and ecosystem health: Past trends and a guide to future projections. Science of the Total Environment 400(1-3):257-269.

Carlson, D.L., I. Basu, and R.A. Hites. 2004. Annual variations of pesticide concentrations in great lakes precipitation. Environmental Science & Technology 38(20):5290-5296.

Carlson, T.N. 1998. Mid-latitude weather systems. Boston: American Meteorological Society.

Carmichael, G.R., T. Sakurai, D. Streets, Y. Hozumi, H. Ueda, S.U. Park, C. Fung, Z. Han, M. Kajino, M. Engardt, C. Bennet, H. Hayami, K. Sartelet, T. Holloway, Z. Wang, A. Kannari, J. Fu, K. Matsuda, N. Thongboonchoo, and M. Amann. 2008a. Mics-Asia ii: The model intercomparison study for Asia phase ii methodology and overview of findings. Atmospheric Environment 42(15):3468-3490.

Carmichael, G.R., A. Sandu, T. Chai, D.N. Daescu, E.M. Constantinescu, and Y. Tang. 2008b. Predicting air quality: Improvements through advanced methods to integrate models and measurements. Journal of Computational Physics 227(7):3540-3571.

Chen, C.Y., N. Serell, D.C. Evers, B.J. Fleishman, K.F. Lambert, J. Weiss, R.P. Mason, and M.S. Bank. 2008. Meeting report: Methylmercury in marine ecosystems—from sources to seafood consumers. Environmental Health Perspectives 116(12):1706-1712.

Chen, Y., and R.G. Prinn. 2006. Estimation of atmospheric methane emissions between 1996 and 2001 using a three-dimensional global chemical transport model. Journal of Geophysical Research 111:D10307, doi:10.1029/ 2005JD006058.

Chin, M., T. Diehl, P. Ginoux, and W. Malm. 2007. Intercontinental transport of pollution and dust aerosols: Implications for regional air quality. Atmospheric Chemistry and Physics 7(21):5501-5517.

Chin, M., D.J. Jacob, J.W. Munger, D.D. Parrish, and B.G. Doddridge. 1994. Relationship of ozone and carbon monoxide over North America. Journal of Geophysical Research 99(D7):14565-14573.

Choi, Y.J., and H.J.S. Fernando. 2007. Simulation of smoke plumes from agricultural burns: Application to the San Luis/Rio Colorado airshed along the U.S./Mexico border. Science of the Total Environment 388(1-3):270-289.

Clarkson, T.W. 2002. The three modern faces of mercury. Environmental Health Perspectives 110(SUPPL. 1):11-23.

Clarkson, T.W., L. Magos, and G.J. Myers. 2003. The toxicology of mercury—current exposures and clinical manifestations. New England Journal of Medicine 349(18):1731-1737.

Cofala, J., M. Amann, Z. Klimont, K. Kupiainen, and L. Höglund-Isaksson. 2007. Scenarios of global anthropogenic emissions of air pollutants and methane until 2030. Atmospheric Environment 41(38):8486-8499.

Cohan, D.S., J. Xu, R. Greenwald, M.H. Bergin, and W.L. Chameides. 2002. Impact of atmospheric aerosol light scattering and absorption on terrestrial net primary productivity. Global Biogeochemical Cycles 16(4):37-1.

Cohen, A.J., H.R. Anderson, B. Ostro, K.D. Pandey, M. Krzyzanowski, N. Kuenzli, K. Gutschmidt, C.A. Pope, I. Romieu, J.M. Samet, and K.R. Smith. 2004. Urban air pollution. In Comparative quantification of health risks: Global and regional burden of disease due to selected major risk factors, edited by M. Ezzati, A.D. Lopez, A. Rodgers and C. J. L. Murray, pp. 1353-1433. Geneva: World Health Organization.

Colarco, P.R., M.R. Schoeberl, B.G. Doddridge, L.T. Marufu, O. Torres, and E.J. Welton. 2004. Transport of smoke from Canadian forest fires to the surface near Washington, D.C.: Injection height, entrainment, and optical properties. Journal of Geophysical Research 109:D06203.

Colette, A., and G. Ancellet. 2005. Impact of vertical transport processes on the tropospheric ozone layering above Europe: Part II: Climatological analysis of the past 30 years. Atmospheric Environment 39(29):5423-5435.

Collins, W.J., R.G. Derwent, B. Garnier, C.E. Johnson, M.G. Sanderson, and D.S. Stevenson. 2003. The effect of stratosphere- troposphere exchange on the future tropospheric ozone trend. Journal of Geophysical Research 108(D12), doi:10.1029/2002JD002617.

Cooper, O.R., C. Forster, D. Parrish, M. Trainer, E. Dunlea, Y. Ryerson, G. Hüber, F. Fehsenfeld, D. Nicks, J. Holloway, J. de Gouw, C. Warneke, J.M. Roberts, F. Flocke, and J. Moody. 2004. A case study transpacific warm conveyor belt transport: Influence of merging airstreams on trace gas import to North America. Journal of Geophysical Research–Atmospheres 109(23):1-17.

Cooper, O.R., J.L. Moody, D.D. Parrish, M. Trainer, J.S. Holloway, G. Hübler, F.C. Fehsenfeld, and A. Stohl. 2002a. Trace gas composition of midlatitude cyclones over the western North Atlantic Ocean: A seasonal comparison of O_3 and CO. Journal of Geophysical Research–Atmospheres 107(7-8):2-1.

Cooper, O.R., J.L. Moody, D.D. Parrish, M. Trainer, T.B. Ryerson, J.S. Holloway, G. Hübler, F.C. Fehsenfeld, and M.J. Evans. 2002b. Trace gas composition of midlatitude cyclones over the western North Atlantic Ocean: A conceptual model. Journal of Geophysical Research–Atmospheres 107(7-8):1-1.

Cooper, O.R., J.L. Moody, D.D. Parrish, M. Trainer, T.B. Ryerson, J.S. Holloway, G. Hübler, F.C. Fehsenfeld, S.J. Oltmans, and M.J. Evans. 2001. Trace gas signatures of the airstreams within North Atlantic cyclones: Case studies from the North Atlantic Regional Experiment (NARE '97) aircraft intensive. Journal of Geophysical Research–Atmospheres 106(D6):5437-5456.

Corbett, J.J., J.J. Winebrake, E.H. Green, P. Kasibhatla, V. Eyring, and A. Lauer. 2007. Mortality from ship emissions: A global assessment. Environmental Science and Technology 41(24):8512-8518.

Cortes, D.R., I. Basu, C.W. Sweet, and R.A. Hites. 2000. Temporal trends in and influence of wind on pah concentrations measured near the great lakes. Environmental Science & Technology 34(3):356-360.

Cotton, W.R., G.D. Alexander, R. Hertenstein, R.L. Walko, R.L. McAnelly, and M. Nicholls. 1995. Cloud venting—a review and some new global annual estimates. Earth Science Reviews 39(3-4):169-206.

Cubison, M.J., B. Ervens, G. Feingold, K.S. Docherty, I.M. Ulbrich, L. Shields, K. Prather, S. Hering, and J.L. Jimenez. 2008. The influence of chemical composition and mixing state of Los Angeles urban aerosol on CCN number and cloud properties. Atmospheric Chemistry and Physics 8(18):5649-5667.

Damoah, R., N. Spichtinger, R. Servranckx, M. Fromm, E.W. Eloranta, I.A. Razenkov, P. James, M. Shulski, C. Forster, and A. Stohl. 2006. A case study of pyro-convection using transport model and remote sensing data. Atmospheric Chemistry and Physics 6(1):173-185.

Daniels, M.J., F. Dominici, J.M. Samet, and S.L. Zeger. 2000. Estimating particulate matter-mortality dose-response curves and threshold levels: An analysis of daily time-series for the 20 largest U.S. cities. American Journal of Epidemiology 152:397-406.

Dalsøren, S.B., M.S. Eide, Ø. Endresen, A. Mjelde, G. Gravir, and I.S.A. Isaksen. 2009. Update on emissions and environmental impacts from the international fleet of ships: The contribution from major ship types and ports. Atmospheric Chemistry and Physics 9(6):2171-2194.

Daly, G.L., and F. Wania. 2004. Simulating the influence of snow on the fate of organic compounds. Environmental Science & Technology 38(15):4176-4186.

Dastoor, A.P., D. Davignon, N. Theys, M. Van Roozendael, A. Steffen, and P.A. Ariya. 2008. Modeling dynamic exchange of gaseous elemental mercury at polar sunrise. Environmental Science & Technology 42(14):5183-5188.

Deacon, R.T., and C.S. Norman. 2006. Does the environmental kuznets curve describe how individual countries behave? Land Economics 82(2):291-315.

DeBell, L.J., K. Gebhart, J. Hand, W. Malm, M. Pitchford, B. Schichtel, and W. White. 2006. Spatial and Seasonal Patterns and Temporal Variability of Haze and its Constituents in the United States, Report IV. Fort Collins, Colo: Cooperative Institute for Research in the Atmosphere.

De Foy, B., J.D. Fast, S.J. Paech, D. Phillips, J.T. Walters, R.L. Coulter, T.J. Martin, M.S. Pekour, W.J. Shaw, P.P. Kastendeuch, N.A. Marley, A. Retama, and L.T. Molina. 2008. Basin-scale wind transport during the milagro field campaign and comparison to climatology using cluster analysis. Atmospheric Chemistry and Physics 8(5):1209-1224.

de Gouw, J.A., and J.L. Jimenez. 2009. Organic aerosols in the Earth's atmosphere. Environmental Science and Technology (in press).

Dentener, F., D. Stevenson, J. Cofala, R. Mechler, M. Amann, P. Bergamaschi, F. Raes, and R. Derwent. 2005. The impact of air pollutant and methane emission controls on tropospheric ozone and radiative forcing: CTM calculations for the period 1990-2030. Atmospheric Chemistry and Physics 7(5):1731-1755.

Dentener, F., S. Kinne, T. Bond, O. Boucher, J. Cofala, S. Generoso, P. Ginoux, S. Gong, J.J. Hoelzemann, A. Ito, L. Marelli, J.E. Penner, J.-P. Putaud, C. Textor, M. Schulz, G.R.v.d. Werf, and J. Wilson. 2006. Emissions of primary aerosol and precursor gases in the years 2000 and 1750: Prescribed data-sets for aerocom. Atmospheric Chemistry and Physics 6:4321-4344.

Derwent, R.G., P.G. Simmonds, A.J. Manning, and T.G. Spain. 2007. Trends over a 20-year period from 1987 to 2007 in surface ozone at the atmospheric research station, Mace Head, Ireland. Atmospheric Environment 41(39):9091-9098.

Derwent, R.G., D.S. Stevenson, W.J. Collins, and C.E. Johnson. 2004. Intercontinental transport and the origins of the ozone observed at surface sites in Europe. Atmospheric Environment 38(13):1891-1901.

Derwent, R.G., D.S. Stevenson, R.M. Doherty, W.J. Collins, and M.G. Sanderson. 2008. How is surface ozone in Europe linked to Asian and North American NO_x emissions? Atmospheric Environment 42(32):7412-7422.

Deshler, T., J.L. Mercer, H.G.J. Smit, R. Stubi, G. Levrat, B.J. Johnson, S.J. Oltmans, R. Kivi, A.M. Thompson, J. Witte, J. Davies, F.J. Schmidlin, G. Brothers, and T. Sasaki. 2008. Atmospheric comparison of electrochemical cell ozonesondes from different manufacturers, and with different cathode solution strengths: The balloon experiment on standards for ozonesondes. Journal of Geophysical Research 113:D04307.

Dickerson, R.R., G.J. Huffman, and W.T. Luke. 1987. Thunderstorms: An important mechanism in the transport of air pollutants. Science 235(4787):460-465.

Dickerson, R.R., C. Li, Z. Li, L.T. Marufu, J.W. Stehr, B. McClure, N. Krotkov, H. Chen, P. Wang, X. Xia, X. Ban, F. Gong, J. Yuan, and J. Yang. 2007. Aircraft observations of dust and pollutants over northeast china: Insight into the meteorological mechanisms of transport. Journal of Geophysical Research 112(D24S90).

Dinglasan, M.J.A., Y. Ye, E.A. Edwards, and S.A. Mabury. 2004. Fluorotelomer alcohol biodegradation yields poly- and perfluorinated acids. Environmental Science & Technology 38(10):2857-2864.

Djuric, D. 1994. Weather Analysis. Englewood Cliffs, NJ: Prentice Hall.

Dlugokencky, E.J., S. Houweling, L. Bruhwiler, K.A. Masarie, P.M. Lang, J.B. Miller, and P.P. Tans. 2003. Atmospheric methane levels off: Temporary pause or a new steady-state? Geophysical Research Letters 30(19).

Dominguez, G., T. Jackson, L. Brothers, B. Barnett, B. Nguyen, and M.H. Thiemens. 2008. Discovery and measurement of an isotopically distinct source of sulfate in earth's atmosphere. Proceedings of the National Academy of Sciences of the United States of America 105(35):12769-12773.

Dommergue, A., C.P. Ferrari, M. Amyot, S. Brooks, F. Sprovieri, and A. Steffen. 2008. Spatial coverage and temporal trends of atmospheric mercury measurements in polar regions. In Mercury fate and transport in the global atmosphere measurements, models and policy implications, edited by N. Pirrone and R. Mason. 220-242. New York: Springer.

Donahue, N., A. Robinson, and S. Pandis. 2009. Atmospheric organic particulate matter: From smoke to secondary organic aerosol. Atmospheric Environment 43(1):94-106.

Donald, D.B., J. Syrgiannis, R.W. Crosley, G. Holdsworth, D.C.G. Muir, B. Rosenberg, A. Sole, and D.W. Schindler. 1999. Delayed deposition of organochlorine pesticides at a temperate glacier. Environmental Science & Technology 33(11):1794-1798.

Donnell, E., D. Fish, E. Dicks, and A. Thorpe. 2001. Mechanisms for pollutant transport between the boundary layer and the free troposphere. Journal of Geophysical Research 106(D8):7847-7856.

Doswell, C.A., ed. 2001. Severe convective storms. Vol. 28, Meteorological monographs. Boston, MA: American Meteorological Society. 561 pp.

Duncan, B.N., and I. Bey. 2004. A modeling study of the export pathways of pollution from Europe: Seasonal and interannual variations (1987-1997). Journal of Geophysical Research–Atmospheres 109(8).

Dvonch, J.T., J.R. Graney, G.J. Keeler, and R.K. Stevens. 1999. Use of elemental tracers to source apportion mercury in south Florida precipitation. Environmental Science and Technology 33(24):4522-4527.

Ebinghaus, R. 2008. Mercury cycling in the Arctic—does enhanced deposition flux mean net-input? Environmental Chemistry 5(2):87-88.

Eckhardt, S., A. Stohl, H. Wernli, P. James, C. Forster, and N. Spichtinger. 2004. A 15-year climatology of warm conveyor belts. Journal of Climate 17(1):218-237.

Edgerton, E.S., B.E. Hartsell, and J.J. Jansen. 2006. Mercury speciation in coal-fired power plant plumes observed at three surface sites in the southeastern U.S. Environmental Science and Technology 40(15):4563-4570.

Edwards, D.P., L.K. Emmons, D.A. Hauglustaine, D.A. Chu, J.C. Gille, Y.J. Kaufman, G. Pétron, L.N. Yurganov, L. Giglio, M.N. Deeter, V. Yudin, D.C. Ziskin, J. Warner, J.F. Lamarque, G.L. Francis, S.P. Ho, D. Mao, J. Chen, E.I. Grechko, and J.R. Drummond. 2004. Observations of carbon monoxide and aerosols from the Terra satellite: Northern hemisphere variability. Journal of Geophysical Research 109:D24202.

Engelstaedter, S., and R. Washington. 2007. Temporal controls on global dust emissions: The role of surface gustiness. Geophysical Research Letters 34(15).

Engle, M.A., M.S. Gustin, F. Goff, D.A. Counce, C.J. Janik, D. Bergfeld, and J.J. Rytuba. 2006. Atmospheric mercury emissions from substrates and fumaroles associated with three hydrothermal systems in the western United States. Journal of Geophysical Research–Atmospheres 111(D17304).

Engstrom, D.R., and E.B. Swain. 1997. Recent declines in atmospheric mercury deposition in the upper midwest. Environmental Science and Technology 31(4):960-967.

EPA. 1997. Mercury study report to congress. Office of Air Quality Planning and Standards and Office of Research and Development, U.S. Environmental Protection Agency. EPA-452/R-97-007. Washington, DC. Available online at http://www.epa.gov/mercury/report. htm.

———. 2004. Air quality criteria for particulate matter (final report, Oct 2004). U.S. Environmental Protection Agency. EPA 600/P-99/002aF-bF. Washington, DC. Available online at http://cfpub.epa.gov/ncea/cfm/recordisplay.cfm?deid=87903.

———. 2006a. Air quality criteria for ozone and related photochemical oxidants (final). U.S. Environmental Protection Agency. EPA/600/R-05/004aF-cF. Washington, DC. Available online at http://purl.access.gpo.gov/GPO/LPS86552.

———. 2006b. Speciate 4.0-speciation database development documentation. November. US Environmental Protection Agency. EPA/600/R-06/161. Washington, DC. Available online at http://www.epa.gov/ttn/chief/software/speciate/index.html.

———. 2009. Particulate matter. U.S. Environmental Protection Agency, Available online at http://epa.gov/air/emissions/pm.htm. [accessed June 09, 2009, 2009].

Ericksen, J.A., M.S. Gustin, S.E. Lindberg, S.D. Olund, and D.P. Krabbenhoft. 2005. Assessing the potential for re-emission of mercury deposited in precipitation from arid soils using a stable isotope. Environmental Science and Technology 39(20):8001-8007.

Esler, J.G., G.J. Roelofs, M.O. Köhler, and F.M. O'Connor. 2004. A quantitative analysis of grid-related systematic errors in oxidising capacity and ozone production rates in chemistry transport models. Atmospheric Chemistry and Physics 4(7):1781-1795.

Evers, D.C., and T.A. Clair. 2005. Mercury in northeastern North America: A synthesis of existing databases. Ecotoxicology 14(1-2):7-14.

Eyring, V., H.W. Köhler, A. Lauer, and B. Lemper. 2005. Emissions from international shipping: 2. Impact of future technologies on scenarios until 2050. Journal of Geophysical Research–Atmospheres 110(D17306):183-200.

Falke, S.R., R.B. Husar, and B.A. Schichtel. 2001. Fusion of seawifs and toms satellite data with surface observations and topographic data during extreme aerosol events. Journal of the Air & Waste Management Association 51(11):1579-1585.

Fast, J.D., B. De Foy, F.A. Rosas, E. Caetano, G. Carmichael, L. Emmons, D. McKenna, M. Mena, W. Skamarock, X. Tie, R.L. Coulter, J.C. Barnard, C. Wiedinmyer, and S. Madronich. 2007. A meteorological overview of the milagro field campaigns. Atmospheric Chemistry and Physics 7(9):2233-2257.

Finlayson-Pitts, B.J., and J.N. Pitts Jr. 2006. Chemistry of the upper and lower atmosphere: Theory, experiments, and applications. San Diego: Academic Press.

Finley, B., P.C. Swartzendruber, and D.A. Jaffe. 2009. Large particulate mercury emissions in regional wildfire plumes observed at the mount bachelor observatory. Atmospheric Environment In Press, Corrected Proof.

Fiore, A., D.J. Jacob, H. Liu, R.M. Yantosca, T.D. Fairlie, and Q. Li. 2003a. Variability in surface ozone background over the United States: Implications for air quality policy. Journal of Geophysical Research–Atmospheres 108(D24).4787.

Fiore, A.M., F.J. Dentener, O. Wild, C. Cuvelier, M.G. Schultz, P. Hess, C. Textor, M. Schulz, R.M. Doherty, L.W. Horowitz, I.A. MacKenzie, M.G. Sanderson, D.T. Shindell, D.S. Stevenson, S. Szopa, R. Van Dingenen, G. Zeng, C. Atherton, D. Bergmann, I. Bey, G. Carmichael, W.J. Collins, B.N. Duncan, G. Faluvegi, G. Folberth, M. Gauss, S. Gong, D. Hauglustaine, T. Holloway, I.S.A. Isaksen, D.J. Jacob, J.E. Jonson, J.W. Kaminski, T.J. Keating, A. Lupu, E. Marmer, V. Montanaro, R.J. Park, G. Pitari, K.J. Pringle, J.A. Pyle, S. Schroeder, M.G. Vivanco, P. Wind, G. Wojcik, S. Wu, and A. Zuber. 2009. Multimodel estimates of intercontinental source-receptor relationships for ozone pollution. Journal of Geophysical Research–Atmospheres 114:D04301.

Fiore A.M., L.W. Horowitz, E.J. Dlugokencky, and J.J. West. 2006. Impact of meteorology and emissions on methane trends. Geophysical Research Letters 33:L12809, doi:10.1029/2006GL026199.

Fiore, A.M., D.J. Jacob, R. Mathur, and R.V. Martin. 2003b. Application of empirical orthogonal functions to evaluate ozone simulations with regional and global models. Journal of Geophysical Research–Atmospheres 108(D14):4431.

Fiore, A.M., J.J. West, L.W. Horowitz, V. Naik, and M.D. Schwarzkopf. 2008. Characterizing the tropospheric ozone response to methane emission controls and the benefits to climate and air quality. Journal of Geophysical Research–Atmospheres 113(D8):D08307.

Fiore, A.M., D.J. Jacob, I. Bey, R.M. Yantosca, B.D. Field, A.C. Fusco, and J.G. Wilkinson. 2002. Background ozone over the United States in summer: Origin, trend, and contribution to pollution episodes. Journal of Geophysical Research–Atmospheres 107(15).

Fischer, E.V., N.C. Hsu, D.A. Jaffe, M.J. Jeong, and S.L. Gong. 2009. A decade of dust: Asian dust and springtime aerosol load in the us Pacific Northwest. Geophysical Research Letters 36.

Fitzgerald, W.F., D.R. Engstrom, C.H. Lamborg, C.M. Tseng, P.H. Balcom, and C.R. Hammerschmidt. 2005. Modern and historic atmospheric mercury fluxes in northern alaska: Global sources and arctic depletion. Environmental Science and Technology 39(2):557-568.

Fitzgerald, W.F., D.R. Engstrom, R.P. Mason, and E.A. Nater. 1998. The case for atmospheric mercury contamination in remote areas. Environmental Science and Technology 32(1):1-7.

Flanner, M.G., C.S. Zender, J.T. Randerson, and P.J. Rasch. 2007. Present-day climate forcing and response from black carbon in snow. Journal of Geophysical Research 112:D11202, doi:10.1029/2006JD008003.

Flannigan, M.D., K.A. Logan, B.D. Amiro, W.R. Skinner, and B.J. Stocks. 2005. Future area burned in Canada. Climatic Change 72(1-2):1-16.

Forster, P., V. Ramaswamy, P. Artaxo, T. Berntsen, R. Betts, D.W. Fahey, J. Haywood, J. Lean, D.C. Lowe, G. Myhre, J. Ng'ang'a, R. Prinn, G. Raga, M. Schulz, and R.V. Dorland. 2007a. Changes in atmospheric constituents and radiative forcing. In Climate change 2007: The physical science basis. Contribution of working group i to the fourth assessment report of the intergovernmental panel on climate change, edited by S. Solomon, D. Qin, M. Manning, M. Marquis, K. Averyt, M. M. B. Tignor and H. L. Miller. 129-234, Cambridge, United Kingdom and New York: Cambridge University Press.

Forster, P.M., G. Bodeker, R. Schofield, S. Solomon, and D. Thompson. 2007b. Effects of ozone cooling in the tropical lower stratosphere and upper troposphere. Geophysical Research Letters 34(23).

Forster, C., U. Wandinger, G. Wotawa, P. James, I. Mattis, D. Althausen, P. Simmonds, S. O'Doherty, S.G. Jennings, C. Kleefeld, J. Schneider, T. Trickl, S. Kreipl, H. Jäger, and A. Stohl. 2001. Transport of boreal forest fire emissions from Canada to Europe. Journal of Geophysical Research–Atmospheres 106(D19):22887-22906.

Friedli, H.R., L.F. Radke, and J.Y. Lu. 2001. Mercury in smoke from biomass fires. Geophysical Research Letters 28(17):3223-3226.

Friedli, H.R., L.F. Radke, J.Y. Lu, C.M. Banic, W.R. Leaitch, and J.I. MacPherson. 2003. Mercury emissions from burning of biomass from temperate North American forests: Laboratory and airborne measurements. Atmospheric Environment 37(2):253-267.

Fromm, M., J. Alfred, K. Hoppel, J. Hornstein, R. Bevilacqua, E. Shettle, R. Servranckx, Z. Li, and B. Stocks. 2000. Observations of boreal forest fire smoke in the stratosphere by poam iii, sage ii, and lidar in 1998. Geophysical Research Letters 27(9):1407-1410.

Fromm, M., R. Bevilacqua, R. Servranckx, J. Rosen, J.P. Thayer, J. Herman, and D. Larko. 2005. Pyro-cumulonimbus injection of smoke to the stratosphere: Observations and impact of a super blowup in northwestern Canada on August 3-4, 1998. Journal of Geophysical Research–Atmospheres 110(8):1-17.

Fu, T.-M., D.J. Jacob, P.I. Palmer, K. Chance, Y.X. Wang, B. Barletta, D.R. Blake, J.C. Stanton, and M.J. Pilling. 2007. Space-based formaldehyde measurements as constraints on volatile organic compound emissions in East and South Asia and implications for ozone. Journal of Geophysical Research 112:D06312.

Furutani, H., M. Dall'osto, G.C. Roberts, and K.A. Prather. 2008. Assessment of the relative importance of atmospheric aging on ccn activity derived from field observations. Atmospheric Environment 42(13):3130-3142.

Fusco, A.C., and J.A. Logan. 2003. Analysis of 1970-1995 trends in tropospheric ozone at Northern Hemisphere midlatitudes with the geos-chem model. Journal of Geophysical Research–Atmospheres 108(15).

Gallagher, K.P., and R. Taylor. 2003. International trade and air pollution: The economic costs of air emissions from waterborne commerce vessels in the United States. September. Tufts University. No. 03-08. Medford. Available online at http://www.ase.tufts.edu/gdae/publications/working_papers/03-08WaterborneCommerce.PDF.

García-Hernández, J., L. Cadena-Cárdenas, M. Betancourt-Lozano, L.M. García-De-La-Parra, L. García-Rico, and F. Márquez-Farías. 2007. Total mercury content found in edible tissues of top predator fish from the Gulf of California, Mexico. Toxicological and Environmental Chemistry 89(3):507-522.

Garrison, V.H., W.T.F., S. Genualdi, D.W. Griffin, C.A. Kellogg, M.S. Majewski, A. Moham-med, A. Ramsubhag, E.A.Shinn, S.L. Simonich, and G.W. Smith. 2006. Saharan dust—a carrier of persistent organic pollutants, metals, and microbes to the Caribbean? Revista de Biologia Tropical–International Journal of Tropical Biology and Conservation 54 (Suppl. 3):9-21.

Gauss, M., G. Myhre, I.S.A. Isaksen, V. Grewe, G. Pitari, O. Wild, W.J. Collins, F.J. Dentener, K. Ellingsen, L.K. Gohar, D.A. Hauglustaine, D. Iachetti, J.F. Lamarque, E. Mancini, L.J. Mickley, M.J. Prather, J.A. Pyle, M.G. Sanderson, K.P. Shine, D.S. Stevenson, K. Sudo, S. Szopa, and G. Zeng. 2006. Radiative forcing since preindustrial times due to ozone change in the troposphere and the lower stratosphere. Atmospheric Chemistry and Physics 6(3):575-599.

Gebhart, K.A., S.M. Kreidenweis, and W.C. Malm. 2001. Back-trajectory analyses of fine particulate matter measured at big bend national park in the historical database and the 1996 scoping study. Science of the Total Environment 276(1-3):185-204.

Geisz, H.N., R.M. Dickhut, M.A. Cochran, W.R. Fraser, and H.W. Ducklow. 2008. Melting glaciers: A probable source of ddt to the antarctic marine ecosystem. Environmental Science & Technology 42(11):3958-3962.

Genualdi, S.A., R.K. Killin, J. Woods, G. Wilson, D. Schmedding, and S.L.M. Simonich. 2009. Trans-pacific and regional atmospheric transport of polycyclic aromatic hydrocarbons and pesticides in biomass burning emissions to western North America. Environmental Science & Technology 43(4):1061-1066.

Genualdi, S.A., S.L. Massey Simonich, T.K. Primbs, T.F. Bidleman, L.M. Jantunen, K.S. Ryoo, T. Zhu. 2009. Enantiomeric signature of organochlorine pesticides in Asian, trans-Pacific and western U.S. air masses. Environmental Science & Technology 43:2806-2811.

Girardin, M.P., and M. Mudelsee. 2008. Past and future changes in Canadian boreal wildfire activity. Ecological Applications 18(2):391-406.

Glassmeyer, S.T., D.S. De Vault, and R.A. Hites. 2000. Rates at which toxaphene concentra-tions decrease in lake trout from the great lakes. Environmental Science & Technology 34(9):1851-1855.

Glassmeyer, S.T., D.S. DeVault, T.R. Myers, and R.A. Hites. 1997. Toxaphene in great lakes fish: A temporal, spatial, and trophic study. Environmental Science & Technology 31(1):84-88.

Goldstein, A.H., M. McKay, M.R. Kurpius, G.W. Schade, A. Lee, R. Holzinger, and R.A. Rasmussen. 2004. Forest thinning experiment confirms ozone deposition to forest canopy is dominated by reaction with biogenic VOCs. Geophysical Research Letters 31(22):1-4.

Grieshop, A.P., J.M. Logue, N.M. Donahue, and A.L. Robinson. 2008. Laboratory investiga-tion of photochemical oxidation of organic aerosol from wood fires-part 1: Measure-ment and simulation of organic aerosol evolution. Atmospheric Chemistry and Physics Discussions 8(4):15699-15737.

Gryparis, A., B. Forsberg, K. Katsouyanni, A. Analitis, G. Touloumi, J. Schwartz, E. Samoli, S. Medina, H.R. Anderson, E.M. Niciu, H.E. Wichmann, B. Kriz, M. Kosnik, J. Skorkovsky, J.M. Vonk, Z. Dortbudak. 2004. Acute effects of ozone on mortality from the "Air Pollution and Health: A European Approach" Project. American Journal of Respiratory and Critical Care Medicine 170:1080-1087.

Grigal, D.F. 2003. Mercury sequestration in forests and peatlands: A review. Journal of Envi-ronmental Quality 32(2):393-405.

Guenther, A., C. Geron, T. Pierce, B. Lamb, P. Harley, and R. Fall. 2000. Natural emissions of non-methane volatile organic compounds, carbon monoxide, and oxides of nitrogen from North America. Atmospheric Environment 34: 2205-2230.

Guenther, A., T. Karl, P. Harley, C. Wiedinmyer, P.I. Palmer, and C. Geron. 2006. Estimates of global terrestrial isoprene emissions using megan (model of emissions of gases and aerosols from nature). Atmospheric Chemistry and Physics 6:3181-3210.

Guentzel, J.L., W.M. Landing, G.A. Gill, and C.D. Pollman. 2001. Processes influencing rainfall deposition of mercury in Florida Environmental Science and Technology 35(5):863-873

Guerova, G., I. Bey, J.-L. Attie, and R.V. Martin. 2006. Case studies of ozone between North America and Europe in summer 2000, Atmospheric Chemistry and Physics Discussion 5:6127-6184.

Guo, Z.G., T. Lin, G. Zhang, Z.S. Yang, and M. Fang. 2006. High-resolution depositional records of polycyclic aromatic hydrocarbons in the central continental shelf mud of the east china sea. Environmental Science & Technology 40(17):5304-5311.

Gurjar, B.R., T.M. Butler, M.G. Lawrence, and J. Lelieveld. 2008. Evaluation of emissions and air quality in megacities. Atmospheric Environment 42(7):1593-1606.

Gusev, A., O. Rozovskaya, and V. Shatalov. 2007. Modelling pop long-range transport and contamination levels by msce-pop model. April. Meteorological Synthesizing Centre–East. EMEP/MSC-E Technical Report 1/2007. Moscow, Russia. Available online at http://www.msceast.org/publications.html.

Gustin, M., and S. Lindberg. 2005. Terrestrial mercury fluxes: Is the net exchange up, down or neither? In Dynamics of mercury pollution on regional and global scales: Atmospheric processes, human exposure around the world, N. Pironne and K. Mahaffey, eds. Nowell, MA: USA Springer Publisher.

Gustin, M.S., S.E. Lindberg, and P.J. Weisberg. 2008. An update on the natural sources and sinks of atmospheric mercury. Applied Geochemistry 23(3):482-493.

Hadley, O.L., V. Ramanathan, G.R. Carmichael, Y. Tang, C.E. Corrigan, G.C. Roberts, and G.S. Mauger. 2007. Trans-pacific transport of black carbon and fine aerosols (d < 2.5 μm) into North America. Journal of Geophysical Research 112:D05309.

Hafner, W.D., D.L. Carlson, and R.A. Hites. 2005. Influence of local human population on atmospheric polycyclic aromatic hydrocarbon concentrations. Environmental Science & Technology 39(19):7374-7379.

Hafner, W.D., and R.A. Hites. 2003. Potential sources pesticides, pcbs, and pahs to the atmosphere of the great lakes. Environmental Science & Technology 37(17):3764-3773.

———. 2005. Effects of wind and air trajectory directions on atmospheric concentrations of persistent organic pollutants near the great lakes. Environmental Science & Technology 39(20):7817-7825.

Hageman, K.J., S.L. Simonich, D.H. Campbell, G.R. Wilson, and D.H. Landers. 2006. Atmospheric deposition of current-use and historic-use pesticides in snow at national parks in the western United States. Environmental Science & Technology 40(10):3174-3180.

Hakami, A., D.K. Henze, J.H. Seinfeld, K. Singh, A. Sandu, S. Kim, D. Byun, and Q. Li. 2007. The adjoint of CMAQ. Environmental Science and Technology 41(22):7807-7817.

Hakami, A., J.H. Seinfeld, T. Chai, Y. Tang, G.R. Carmichael, and A. Sandu. 2006. Adjoint sensitivity analysis of ozone nonattainment over the continental United States. Environmental Science and Technology 40(12):3855-3864.

Halland, J.J., H.E. Fuelberg, K.E. Pickering, and M. Luo. 2009. Identifying convective transport of carbon monoxide by comparing remotely sensed observations from TES with cloud modeling simulations. Atmospheric Chemistry and Physics 9:4279-4294.

Halsall, C.J. 2004. Investigating the occurrence of persistent organic pollutants (pops) in the arctic: Their atmospheric behaviour and interaction with the seasonal snow pack. Environmental Pollution 128(1-2):163-175.

Hammerschmidt, C.R., and W.F. Fitzgerald. 2006. Methylmercury in freshwater fish linked to atmospheric mercury deposition. Environmental Science and Technology 40(24):7764-7770.

Hammerschmidt, C.R., C.H. Lamborg, and W.F. Fitzgerald. 2007. Aqueous phase methylation as a potential source of methylmercury in wet deposition. Atmospheric Environment 41(8):1663-1668.

Harris, N.R.P., G. Ancellet, L. Bishop, D.J. Hofmann, J.B. Kerr, R.D. McPeters, M. Prendez, W.J. Randel, J. Staehelin, B.H. Subbaraya, A. Volz-Thomas, J. Zawodny, and C.S. Zerefos. 1997. Trends in stratospheric and free tropospheric ozone. Journal of Geophysical Research–Atmospheres 102(1):1571-1590.

Harris, R.C., J.W.M. Rudd, M. Amyot, C.L. Babiarz, K.G. Beaty, P.J. Blanchfield, R.A. Bodaly, B.A. Branfireun, C.C. Gilmour, J.A. Graydon, A. Heyes, H. Hintelmann, J.P. Hurley, C.A. Kelly, D.P. Krabbenhoft, S.E. Lindberg, R.P. Mason, M.J. Paterson, C.L. Podemski, A. Robinson, K.A. Sandilands, G.R. Southworthn, V.L. St. Louis, and M.T. Tate. 2007. Whole-ecosystem study shows rapid fish-mercury response to changes in mercury deposition. Proceedings of the National Academy of Sciences of the United States of America 104(42):16586-16591.

Hartman, J., P. Weisberg, R. Pillai, J.A. Ericksen, T. Kuiken, S. Lindberg, H. Zhang, J. Rytuba, and M.S. Gustin. 2009. Application of a rule-based model to estimate mercury exchange for three background biomes in the continental United States. Environmental Science and Technology 43:4989-4994.

He, J.Z., K.R. Robrock, and L. Alvarez-Cohen. 2006. Microbial reductive debromination of polybrominated diphenyl ethers (pbdes). Environmental Science & Technology 40(14):4429-4434.

Heald, C.L. and D.V. Spracklen. 2009. Atmospheric budget of primary biological aerosol particles from fungal spores. Geophysical Research Letters 36:L09806, doi:10.1029/2009GL037493.

Heald, C.L., D.J. Jacob, R.J. Park, L.M. Russell, B.J. Huebert, J.H. Seinfeld, H. Liao, and R.J. Weber. 2005. A large organic aerosol source in the free troposphere missing from current models. Geophysical Research Letters 32(18):1-4.

Heald, C.L., D.K. Henze, L.W. Horowitz, J. Feddema, J.-F. Lamarque, A. Guenther, P.G. Hess, F. Vitt, J.H. Seinfeld, A.H. Goldstein, and I. Fung. 2008. Predicted change in global secondary organic aerosol concentrations in response to future climate, emissions, and land use change. Journal of Geophysical Research 113:D05211, doi:10.1029/2007JD009092.

Heald, C.L., D.J. Jacob, S. Turquety, R.C. Hudman, R.J. Weber, A.P. Sullivan, R.E. Peltier, E.L. Atlas, J.A. de Gouw, C. Warneke, J.S. Holloway, J.A. Neuman, F.M. Flocke, and J.H. Seinfeld. 2006. Concentrations and sources of organic carbon aerosols in the free troposphere over North America. Journal of Geophysical Research 111:D23S47.

Heald, C.L., D.J. Jacob, A.M. Fiore, L.K. Emmons, J.C. Gille, M.N. Deeter, J. Warner, D.P. Edwards, J.H. Crawford, A.J. Hamlin, G.W. Sachse, E.V. Browell, M.A. Avery, S.A. Vay, D.J. Westberg, D.R. Blake, H.B. Singh, S.T. Sandholm, R.W. Talbot, and H.E. Fuelberg. 2003. Asian outflow and trans-Pacific transport of carbon monoxide and ozone pollution: An integrated satellite, aircraft, and model perspective. Journal of Geophysical Research–Atmospheres 108(D24):4804.

Hedgecock, I.M., and N. Pirrone. 2004. Chasing quicksilver: Modeling the atmospheric lifetime of Hg 0(g) in the marine boundary layer at various latitudes. Environmental Science and Technology 38(1):69-76.

Helmig, D., J. Arey, W.P. Harger, R. Atkinson, and J. Lopezcancio. 1992. Formation of mutagenic nitrodibenzopyranones and their occurrence in ambient air. Environmental Science & Technology 26(3):622-624.

Henne, S., M. Furger, S. Nyeki, M. Steinbacher, B. Neininger, S.F.J. de Wekker, J. Dommen, N. Spichtinger, A. Stohl, and A.S.H. Prévôt. 2004. Quantification of topographic venting of boundary layer air to the free troposphere. Atmospheric Chemistry and Physics 4(2):497-509.

Henze, D.K., J.H. Seinfeld, and D.T. Shindell. 2008. Inverse modeling and mapping us air qual-ity influences of inorganic $PM_{2.5}$ precursor emissions using the adjoint of GEOS-chem. Atmospheric Chemistry and Physics Discussions 8(4):15031-15099.

Hightower, J.M., and D. Moore. 2003. Mercury levels in high-end consumers of fish. Envi-ronmental Health Perspectives 111(4):604-608.

Hillery, B.R., M.F. Simcik, I. Basu, R.M. Hoff, W.M.J. Strachan, D. Burniston, C.H. Chan, K.A. Brice, C.W. Sweet, and R.A. Hites. 1998. Atmospheric deposition of toxic pollut-ants to the great lakes as measured by the integrated atmospheric deposition network. Environmental Science & Technology 32(15):2216-2221.

Hintelmann, H., P. Dillon, R. Douglas Evans, J.W.M. Rudd, R.A. Bodaly, and T.A. Jackson. 2001. Comment: Variations in the isotope composition of mercury in a freshwater sediment sequence and food web1. Canadian Journal of Fisheries and Aquatic Sciences 58(11):2309-2311.

Hintelmann, H., and R.D. Evans. 1997. Application of stable isotopes in environmental tracer studies—measurement of monomethylmercury (CH_3Hg^+) by isotope dilution ICP-MS and detection of species transformation. Fresenius' Journal of Analytical Chemistry 358(3):378-385.

Hintelmann, H., R. Harris, A. Heyes, J.P. Hurley, C.A. Kelly, D.P. Krabbenhoft, S. Lindberg, J.W.M. Rudd, K.J. Scott, and V.L. St. Louis. 2002. Reactivity and mobility of new and old mercury deposition in a boreal forest ecosystem during the first year of the metaalicus study. Environmental Science and Technology 36(23):5034-5040.

Hites, R.A. 1999. Temperature dependence and temporal trends of polychlorinated biphenyl congeners in the great lakes atmosphere. Abstracts of Papers of the American Chemical Society 217:U741-U741.

Hogrefe, C., B. Lynn, K. Civerolo, J.-Y. Ku, J. Rosenthal, C. Rosenzweig, R. Goldberg, S. Gaffin, K. Knowlton, and P.L. Kinney. 2004. Simulating changes in regional air pollution over the eastern United States due to changes in global and regional climate and emis-sions. Journal of Geophysical Research 109:D22301, doi:10.1029/2004JD004690.

Holzer, M., T.M. Hall, and R.B. Stull. 2005. Seasonality and weather-driven variability of transpacific transport. Journal of Geophysical Research–Atmospheres 110(23):1-22.

Houze, R.A. Jr. 1989: Observed structure of mesoscale convective systems and implications for large-scale heating. Quarterly Journal of the Royal Meterological Society 115:425-461.

Howe, T.S., S. Billings, and R.J. Stolzberg. 2004. Sources of polycyclic aromatic hydrocarbons and hexachlorobenzene in spruce needles of eastern Alaska. Environmental Science & Technology 38(12):3294-3298.

HTAP-TF. 2007. Hemispheric transport of air pollution 2007. T. Keating, A. Zuber, G. Lough, O. Cooper, A. Stohl, D. Parrish, Z. Klimont, D. Streets, G. Carmichael and F. Dentener, eds. Geneva: UN–Economic Commission for Europe. Air Pollution Studies No. 16:145.

Hsu, J., and M.J. Prather. 2009. Stratospheric variability and tropospheric ozone. Journal of Geophysical Research 114:D06102, doi:10.1029/2008JD010942.

Hua, I., C. Jafvert, J. Bezares-Cruz, N. Kang, and T. Filley. 2003. Photochemical reaction rates and products of solar irradiated polybrominated diphenylethers. Abstracts of Papers of the American Chemical Society 226:U511-U511.

Hubbell, B.J., A. Hallberg, D.R. McCubbin, and E. Post. 2005. Health-related benefits of attaining the 8-hr ozone standard. Environmental Health Perspectives 113(1):73-82.

Hudman, R.C., D.J. Jacob, S. Turquety, E.M. Leibensperger, L.T. Murray, S. Wu, A.B. Gilliland, M. Avery, T.H. Bertram, W. Brune, R.C. Cohen, J.E. Dibb, F.M. Flocke, A. Fried, J. Holloway, J.A. Neuman, R. Orville, A. Perring, X. Ren, G.W. Sachse, H.B. Singh, A. Swanson, and P.J. Wooldridge. 2006. Surface and lightning sources of nitrogen oxides over the United States: magnitudes, chemical evolution, and outflow. Journal of Geophysical Research 112:D12S05, doi:10.1029/2006JD007912.

Hudman, R.C., D.J. Jacob, O.R. Cooper, M.J. Evans, C.L. Heald, R.J. Park, F. Fehsenfeld, F. Flocke, J. Holloway, G. Hubler, K. Kita, M. Koike, Y. Kondo, A. Neuman, J. Nowak, S. Oltmans, D. Parrish, J.M. Roberts, and T. Ryerson. 2004. Ozone production in transpacific Asian pollution plumes and implications for ozone air quality in California. Journal of Geophysical Research–Atmospheres 109(23):1-14.

Hudman, R.C., L.T. Murray, D.J. Jacob, D.B. Millet, S. Turquety, S. Wu, D.R. Blake, A.H. Goldstein, J. Holloway, and G.W. Sachse. 2008. Biogenic versus anthropogenic sources of CO in the United States. Geophysical Research Letters 35:L04801.

Huebert, B.J., T. Bates, P.B. Russell, G. Shi, Y.J. Kim, K. Kawamura, G. Carmichael, and T. Nakajima. 2003. An overview of ace-Asia: Strategies for quantifying the relationships between Asian aerosols and their climatic impacts. Journal of Geophysical Research–Atmospheres 108(23).

Hung, H., P. Blanchard, G. Poole, B. Thibert, and C.H. Chiu. 2002a. Measurement of particle-bound polychlorinated dibenzo-p-dioxins and dibenzofurans (pcdd/fs) in arctic air at alert, Nunavut, Canada. Atmospheric Environment 36(6):1041-1050.

Hung, H., C.J. Halsall, P. Blanchard, H.H. Li, P. Fellin, G. Stern, and B. Rosenberg. 2001. Are pcbs in the Canadian arctic atmosphere declining? Evidence from 5 years of monitoring. Environmental Science & Technology 35(7).1303-1311.

————. 2002b. Temporal trends of organochlorine pesticides in the Canadian arctic atmosphere. Environmental Science & Technology 36(5):862-868.

Huntrieser, H., J. Heland, H. Schlager, C. Forster, A. Stohl, H. Aufmhoff, F. Arnold, H.E. Scheel, M. Campana, S. Gilge, R. Eixmann, and O. Cooper. 2005. Intercontinental air pollution transport from North America to Europe: Experimental evidence from airborne measurements and surface observations. Journal of Geophysical Research–Atmospheres 110(D01305)

Husar, R.B., D.M. Tratt, B.A. Schichtel, S.R. Falke, F. Li, D. Jaffe, S. Gasso, T. Gill, N.S. Laulainen, F. Lu, M.C. Reheis, Y. Chun, D. Westphal, B.N. Holben, C. Gueymard, I. McKendry, N. Kuring, G.C. Feldman, C. McClain, R.J. Frouin, J. Merrill, D. DuBois, F. Vignola, T. Murayama, S. Nickovic, W.E. Wilson, K. Sassen, N. Sugimoto, and W.C. Malm. 2001. Asian dust events of april 1998. Journal of Geophysical Research–Atmospheres 106(D16):18317 18330.

Hynes, A.J., D.L. Donohoue, I.M. Hedgecock, and M.E. Goodsite. 2008. Our current understanding of major chemical and physical processes affecting mercury dynamics in the atmosphere and at the air-water/terrestrial interfaces. In Mercury fate and transport in the global atmosphere measurements, models and policy implications, edited by N. Pirrone and R. Mason. 322-341. New York: Springer.

In, H.J., D.W. Byun, R.J. Park, N.K. Moon, S. Kim, and S. Zhong. 2007. Impact of transboundary transport of carbonaceous aerosols on the regional air quality in the United States: A case study of the South American wildland fire of may 1998. Journal of Geophysical Research–Atmospheres 112(D07201).

IPCC, ed. 1999. Aviation and the global atmosphere : A special report of ipcc working groups i and iii in collaboration with the scientific assessment panel to the montreal protocol on substances that deplete the ozone layer. Edited by J. E. Penner, D. H. Lister, D. J. Griggs, D. J. Dokken and M. McFarland. Cambridge, UK: Cambridge University Press.

————, ed. 2007. Climate change 2007: The physical science basis. Contribution of working group i to the fourth assessment report of the intergovernmental panel on climate change. Edited by S. Solomon, D. Qin, M. Manning, Z. Chen, M. Marquis, K. Averyt, M. M. B. Tignor and J. Henry LeRoy Miller. Cambridge, United Kingdom and New York: Cambridge University Press. 996 pp.

IPCS-WHO. 1990. Methylmercury. Environmental health criteria 101. International Programme on Chemical Safety, World Health Organization. Geneva. Available online at http://www.inchem.org/documents/ehc/ehc/ehc101.htm.

Irie, H., K. Sudo, H. Akimoto, A. Richter, J.P. Burrows, T. Wagner, M. Wenig, S. Beirle, Y. Kondo, V.P. Sinyakov, and F. Goutail. 2005. Evaluation of long-term tropospheric NO_2 data obtained by GOME over East Asia in 1996-2002. Geophysical Research Letters 32(11):1-4.

Ito, K., S.F. De Leon, M. Lippmann. 2005. Associations between ozone and daily mortality: analysis and meta-analysis. Epidemiology 16:446-457.

Jacob, D.J., and D.A. Winner. 2009. Effect of climate change on air quality. Atmospheric Environment 43:51–63.

Jacob, D.J., J.A. Logan, and P.P. Murti. 1999. Effect of rising Asian emissions on surface ozone in the United States. Geophysical Research Letters 26(14):2175-2178.

Jacob, D.J., J.A. Logan, G.M. Gardner, R.M. Yevich, C.M. Spivakovsky, S.C. Wofsy, S. Sillman, and M.J. Prather. 1993. Factors regulating ozone over the United States and its export to the global atmosphere. Journal of Geophysical Research 98(D8):14817-14826.

Jaeglé, L., D.A. Jaffe, H.U. Price, P. Weiss-Penzias, P.I. Palmer, M.J. Evans, D.J. Jacob, and I. Bey. 2003. Sources and budgets for co and o3 in the northeastern pacific during the spring of 2001: Results from the phobea-ii experiment. Journal of Geophysical Research–Atmospheres 108(20).

Jaeglé, L., D.J. Jacob, S.A. Strode, and N.E. Selin. 2008. The geos-chem model. In Mercury fate and transport in the global atmosphere measurements, models and policy implications, edited by N. Pirrone and R. Mason. 401-410. New York: Springer.

Jaffe, D., T. Anderson, D. Covert, R. Kotchenruther, B. Trost, J. Danielson, W. Simpson, T. Berntsen, S. Karlsdottir, D. Blake, J. Harris, G. Carmichael, and I. Uno. 1999. Transport of Asian air pollution to North America. Geophysical Research Letters 26(6):711-714.

Jaffe, D., I. Bertschi, L. Jaeglé, P. Novelli, J.S. Reid, H. Tanimoto, R. Vingarzan, and D.L. Westphal. 2004. Long-range transport of siberian biomass burning emissions and impact on surface ozone in western North America. Geophysical Research Letters 31(16).

Jaffe, D., D. Chand, W. Hafner, A. Westerling, and D. Spracklen. 2008. Influence of fires on O_3 concentrations in the western U.S. Environmental Science & Technology 42(16):5885-5891.

Jaffe, D., I. McKendry, T. Anderson, and H. Price. 2003a. Six "new" episodes of trans-Pacific transport of air pollutants. Atmospheric Environment 37(3):391-404.

Jaffe, D., E. Prestbo, P. Swartzendruber, P. Weiss-Penzias, S. Kato, A. Takami, S. Hatakeyama, and Y. Kajii. 2005a. Export of atmospheric mercury from Asia. Atmospheric Environment 39(17):3029-3038.

Jaffe, D., H. Price, D. Parrish, A. Goldstein, and J. Harris. 2003b. Increasing background ozone during spring on the west coast of North America. Geophysical Research Letters 30(12):15-1.

Jaffe, D., and J. Ray. 2007. Increase in surface ozone at rural sites in the western US. Atmospheric Environment 41(26):5452-5463.

Jaffe, D., J. Snow, and O. Cooper. 2003c. The 2001 Asian dust events: Transport and impact on surface aerosol concentrations in the U.S. Eos, Transactions of the AGU 84(46).

Jaffe, D., S. Tamura, and J. Harris. 2005b. Seasonal cycle and composition of background fine particles along the west coast of the U.S. Atmospheric Environment 39(2):297-306.

James, R.R., and R.A. Hites. 2002. Atmospheric transport of toxaphene from the southern United States to the great lakes region. Environmental Science & Technology 36(16):3474-3481.

Jenkin, M.E. 2008. Trends in ozone concentration distributions in the UK since 1990: Local, regional and global influences. Atmospheric Environment 42(21):5434-5445.

Jerrett, M., R.T. Burnett, C.A. Pope III, I. Kazuhiko, G. Thurston, D. Krewski, Y. Shi, E. Calle, and M. Thun. 2009. Long-Term Ozone Exposure and Mortality. New England Journal of Medicine 360:1085-1095.

Jorgenson, J.L. 2001. Aldrin and dieldrin: A review of research on their production, environmental deposition and fate, bioaccumulation, toxicology and epidemiology in the United States. Environmental Health Perspectives 109:113-139.

Jost, H.J., K. Drdla, A. Stohl, L. Pfister, M. Loewenstein, J.P. Lopez, P.K. Hudson, D.M. Murphy, D.J. Cziczo, M. Fromm, T.P. Bui, J. Dean-Day, C. Gerbig, M.J. Mahoney, E.C. Richard, N. Spichtinger, J.V. Pittman, E.M. Weinstock, J.C. Wilson, and I. Xueref. 2004. In situ observations of mid-latitude forest fire plumes deep in the stratosphere. Geophysical Research Letters 31(11).

Kaiser J. 2005. Epidemiology. Mounting evidence indicts fine-particle pollution. Science 307:1858-1861.

Kalashnikova, O.V., and R.A. Kahn. 2008. Mineral dust plume evolution over the atlantic from misr and modis aerosol retrievals. Journal of Geophysical Research 113:D24204, doi:10.1029/2008JD010083.

Kanakidou, M., J.H. Seinfeld, S.N. Pandis, I. Barnes, F.J. Dentener, M.C. Facchini, R. Van Dingenen, B. Ervens, A. Nenes, C.J. Nielsen, F. Swietlicki, J. P. Putaud, Y. Balkanski, S. Fuzzi, J. Horth, G.K. Moortgat, R. Winterhalter, C.E.L. Myhre, K. Tsigaridis, E. Vignati, E. G. Stephanou, and J.Wilson. 2005. Organic aerosol and global climate modelling: A review. Atmospheric Chemistry and Physics 5:1053-1123.

Kannan, K., S.H. Yun, and T.J. Evans. 2005. Chlorinated, brominated, and perfluorinated contaminants in livers of polar bears from Alaska. Environmental Science & Technology 39(23):9057-9063.

Kasibhatla, P., A. Arellano, J.A. Logan, P.I. Palmer, and P. Novelli. 2002. Top-down estimate of a large source of atmospheric carbon monoxide associated with fuel combustion in Asia. Geophysical Research Letters 29(19):doi:10.1029/2002GL015581.

Kasischke, E.S., and M.R. Turetsky. 2006. Recent changes in the fire regime across the North American boreal region—spatial and temporal patterns of burning across Canada and alaska. Geophysical Research Letters 33(9):L09703.

Keeler, G.J., N. Pirrone, R. Bullock, and S. Sillman. 2009. The need for a coordinated global mercury monitoring network for global and regional models validations In Mercury fate and transport in the global atmosphere measurements, models and policy implications, edited by N. Pirrone and R. Mason, pp. 391-426. New York: Springer.

Keeler, G.J., M.S. Landis, G.A. Norris, E.M. Christianson, and J.T. Dvonch. 2006. Sources of mercury wet deposition in eastern Ohio, USA. Environmental Science and Technology 40(19):5874-5881.

Kellerhals, M., S. Beauchamp, W. Belzer, P. Blanchard, F. Froude, B. Harvey, K. McDonald, M. Pilote, L. Poissant, K. Puckett, B. Schroeder, A. Steffen, and R. Tordon. 2003. Temporal and spatial variability of total gaseous mercury in Canada: Results from the Canadian atmospheric mercury measurement network (camnet). Atmospheric Environment 37(7):1003-1011.

Kemper, C., P. Gibbs, D. Obendorf, S. Marvanek, and C. Lenghaus. 1994. A review of heavy metal and organochlorine levels in marine mammals in Australia. Science of the Total Environment 154(2-3):129-139.

Kiley, C.M., and H.E. Fuelberg. 2006. An examination of summertime cyclone transport processes during intercontinental chemical transport experiment (intex-a). Journal of Geophysical Research–Atmospheres 111(24).

Killin, R.K., S.L. Simonich, D.A. Jaffe, C.L. DeForest, and G.R. Wilson. 2004. Transpacific and regional atmospheric transport of anthropogenic semivolatile organic compounds to cheeka peak observatory during the spring of 2002. Journal of Geophysical Research–Atmospheres 109(D23).

Kim, B.M., S. Teffera, and M.D. Zeldin. 2000. Characterization of PM2.5 and PM10 in the South Coast Air Basin of Southern California: Part 1–spatial variations. Journal of the Air and Waste Management Association 50(12):2034-2044.

Kinne, S., M. Schulz, C. Textor, S. Guibert, Y. Balkanski, S.E. Bauer, T. Berntsen, T.F. Berglen, O. Boucher, M. Chin, W. Collins, F. Dentener, T. Diehl, R. Easter, J. Feichter, D. Fillmore, S. Ghan, P. Ginoux, S. Gong, A. Grini, J. Hendricks, M. Herzog, L. Horowitz, I. Isaksen, T. Iversen, A. Kirkevåg, S. Kloster, D. Koch, J.E. Kristjansson, M. Krol, A. Lauer, J.F. Lamarque, G. Lesins, X. Liu, U. Lohmann, V. Montanaro, G. Myhre, J.E. Penner, G. Pitari, S. Reddy, O. Seland, P. Stier, T. Takemura, and X. Tie. 2006. An AEROCOM initial assessment—optical properties in aerosol component modules of global models. Atmospheric Chemistry and Physics 6(7):1815-1834.

Kirk, J.L., V.L. St. Louis, and M.J. Sharp. 2006. Rapid reduction and reemission of mercury deposited into snowpacks during atmospheric mercury depletion events at Churchill, Manitoba, Canada. Environmental Science and Technology 40(24):7590-7596.

Kirkevåg, A., T. Iversen, Ø. Seland, J.B. Debernard, T. Storelvmo, and J.E. Kristjánsson. 2008. On the additivity of climate response to anthropogenic aerosols and CO_2, and the enhancement of future global warming by carbonaceous aerosols. Tellus, Series A: Dynamic Meteorology and Oceanography 60 A(3):513-527.

Kley, D., A. Volz, and F. Mulheims. 1988. Ozone measurements in historic perspective. In Tropospheric ozone: Regional and global scale interactions, edited by I. S. A. Isaksen, Pp. 63-72. Norwell, Mass: D. Reidel.

Klonecki, A., P. Hess, L. Emmons, L. Smith, J. Orlando, and D. Blake. 2003. Seasonal changes in the transport of pollutants into the arctic troposphere-model study. Journal of Geophysical Research–Atmospheres 108(4):15-1.

Koch, D., T.C. Bond, D. Streets, N. Unger, and G.R. van der Werf. 2007. Global impacts of aerosols from particular source regions and sectors. Journal of Geophysical Research–Atmospheres 112(D02205).

Kopacz, M., D.J. Jacob, D.K. Henze, C.L. Heald, D.G. Streets, and Q. Zhang. 2009. Comparison of adjoint and analytical Bayesian inversion methods for constraining Asian sources of carbon monoxide using satellite (MOPITT) measurements of CO columns. Journal of Geophysical Research–Atmospheres 114(D04305).

Kotchenruther, R.A., D.A. Jaffe, H.J. Beine, T.L. Anderson, J.W. Bottenheim, J.M. Harris, D.R. Blake, and R. Schmitt. 2001. Observations of ozone and related species in the northeast pacific during the PHOBEA campaigns 2. Airborne observations. Journal of Geophysical Research–Atmospheres 106(D7):7463-7483.

Kreidenweis, S.M., L.A. Remer, R. Bruintjes, and O. Dubovik. 2001. Smoke aerosol from biomass burning in Mexico: Hygroscopic smoke optical model. Journal of Geophysical Research 106(D5):4831-4844.

Krewski, D. M. Jerrett, R.T. Burnett, R. Ma, E. Hughes, Y. Shi, M.C. Turner, C. Arden Pope III, G. Thurston, E.E. Calle, and M.J. Thun. 2009. Extended Analysis of the American Cancer Society Study of Particulate Air Pollution and Mortality. HEI Report 140 Available online at http://pubs.healtheffects.org/view.php?id=315, accessed August 20, 2009.

Kucklick, J.R., M.M. Krahn, P.R. Becker, B.J. Porter, M.M. Schantz, G.S. York, T.M. O'Hara, and S.A. Wise. 2006. Persistent organic pollutants in Alaskan ringed seal (phoca hispida) and walrus (odobenus rosmarus) blubber. Journal of Environmental Monitoring 8(8):848-854.

Kucklick, J.R., W.D.J. Struntz, P.R. Becker, G.W. York, T.M. O'Hara, and J.E. Bohonowych. 2002. Persistent organochlorine pollutants in ringed seals and polar bears collected from northern Alaska. Science of the Total Environment 287(1-2):45-59.

Künzli, N., R. Kaiser, S. Medina, M. Studnicka, O. Chanel, P. Filliger, M. Herry, F. Horak Jr, V. Puybonnieux-Texier, P. Quénel, J. Schneider, R. Seethaler, J.C. Vergnaud, and H. Sommer. 2000. Public-health impact of outdoor and traffic-related air pollution: A European assessment. Lancet 356(9232):795-801.

Kupiainen, K., and Z. Klimont. 2007. Primary emissions of fine carbonaceous particles in Europe. Atmospheric Environment 41:2156-2170.

Kurokawa, J., K. Yumimoto, I. Uno, and T. Ohara. 2009. Adjoint inverse modeling of nox emissions over eastern china using satellite observations of NO_2 vertical column densities. Atmospheric Environment 43(11):1878-1887.

Lack, D.A., J.J. Corbett, T. Onasch, B. Lerner, P. Massoli, P.K. Quinn, T.S. Bates, D.S. Covert, D. Coffman, B. Sierau, S. Herndon, J. Allan, T. Baynard, E. Lovejoy, A.R. Ravishankara, and E. Williams. 2009. Particulate emissions from commercial shipping: Chemical, physical, and optical properties. Journal of Geophysical Research–Atmospheres 114(D00F04).

Lafreniere, M.J., J.M. Blais, M.J. Sharp, and D.W. Schindler. 2006. Organochlorine pesticide and polychlorinated biphenyl concentrations in snow, snowmelt, and runoff at Bow Lake, Alberta. Environmental Science & Technology 40(16):4909-4915.

Lamborg, C.H., W.F. Fitzgerald, A.W.H. Damman, J.M. Benoit, P.H. Balcom, and D.R. Engstrom. 2002a. Modern and historic atmospheric mercury fluxes in both hemispheres: Global and regional mercury cycling implications. Global Biogeochemical Cycles 16(4):51-1.

Lamborg, C.H., W.F. Fitzgerald, J. O'Donnell, and T. Torgersen. 2002b. A non-steady-state compartment model of global-scale mercury biogeochemistry with interhemispheric atmospheric gradients. Geochimica et Cosmochimica Acta 66(7):1105-1118.

Landers, D.H., C. Gubala, M. Verta, M. Lucotte, K. Johansson, T. Vlasova, and W.L. Lockhart. 1998. Using lake sediment mercury flux ratios to evaluate the regional and continental dimensions of mercury deposition in arctic and boreal ecosystems. Atmospheric Environment 32(5):919-928.

Landers, D.H., S.L. Simonich, D. Jaffe, L. Geiser, D.H. Campbell, A. Schwindt, C. Schreck, M. Kent, W. Hafner, H.E. Taylor, K. Hageman, S. Usenko, L. Ackerman, J. Schrlau, N. Rose, T. Blett, and M.M. Erway. 2008. The fate, transport, and ecological impacts of airborne contaminants in western national parks (USA). U.S. Environmental Protection Agency, Office of Research and Development, NHEERL. Corvallis, Oregon.

Landis, M.S., M.M. Lynam, and R.K. Stevens. 2005. The monitoring and modeling of hg species in support of local, regional and global modeling. In Dynamics of mercury pollution on regional and global scales: Atmospheric processes and human exposures around the world, N. Pirrone and K. R. Mahaffey, eds. 123-151. New York: Springer.

Lang, C., S. Tao, W.X. Liu, Y.X. Zhang, and S. Simonich. 2008. Atmospheric transport and outflow of polycyclic aromatic hydrocarbons from china. Environmental Science & Technology 42(14):5196-5201.

Laurier, F.J.G., R.P. Mason, L. Whalin, and S. Kato. 2003. Reactive gaseous mercury formation in the North Pacific Ocean's marine boundary layer: A potential role of halogen chemistry. Journal of Geophysical Research–Atmospheres 108(D17):4529.

Law, K.S., and A. Stohl. 2007. Arctic air pollution: Origins and impacts. Science 315(5818):1537-1540.

Leaner J., J. Dabrowski, R. Mason, T. Resane, M. Richardson, M. Ginster, R. Euripides, and E. Masekoameng. 2008. Mercury emissions from point sources in South Africa. In Mercury Fate and Transport in the Global Atmosphere: Measurements, models and policy implications, edited by N. Pirrone and R. Mason, pp. 113-130. New York: Springer.

Lee, R.G.M., N.J.L. Green, R. Lohmann, and K.C. Jones. 1999. Seasonal, anthropogenic. Air mass, and meteorological influences on the atmospheric concentrations of polychlorinated dibenzo-p-dioxins and dibenzofurans (PCDD/Fs): Evidence for the importance of diffuse combustion sources. Environmental Science & Technology 33(17):2864-2871.

Lee, R.G.M., G.O. Thomas, and K.C. Jones. 2004. PBDEs in the atmosphere of three locations in Western Europe. Environmental Science & Technology 38(3):699-706.

Lefohn, A.S., S.J. Oltmans, T. Dann, and H.B. Singh. 2001. Present-day variability of background ozone in the lower troposphere. Journal of Geophysical Research–Atmospheres 106(D9):9945-9958.

Lei, W., B. De Foy, M. Zavala, R. Volkamer, and L.T. Molina. 2007. Characterizing ozone production in the Mexico City metropolitan area: A case study using a chemical transport model. Atmospheric Chemistry and Physics 7(5):1347-1366.

Lei, Y.D., and F. Wania. 2004. Is rain or snow a more efficient scavenger of organic chemicals? Atmospheric Environment 38(22):3557-3571.

Leibensperger, E.M. , L.J. Mickley, and D.J. Jacob. 2008. Sensitivity of U.S. air quality to mid-latitude cyclone frequency and implications of 1980-2006 climate change. Atmospheric Chemistry and Physics 8:7075-7086.

Lelieveld, J., H. Berresheim, S. Borrmann, P.J. Crutzen, F.J. Dentener, H. Fischer, J. Feichter, P.J. Flatau, J. Heland, R. Holzinger, R. Korrmann, M.G. Lawrence, Z. Levin, K.M. Markowicz, N. Mihalopoulos, A. Minikin, V. Ramanathan, M. De Reus, G.J. Roelofs, H.A. Scheeren, J. Sciare, H. Schlager, M. Schultz, P. Siegmund, B. Steil, E.G. Stephanou, P. Stier, M. Traub, C. Warneke, J. Williams, and H. Ziereis. 2002. Global air pollution crossroads over the mediterranean. Science 298(5594):794-799.

Lelieveld, J., and P.J. Crutzen. 1994. Role of deep cloud convection in the ozone budget of the troposphere. Science 264(5166):1759-1761.

Levy, H., II, M.D. Schwarzkopf, L. Horowitz, V. Ramaswamy, and K.L. Findell. 2008. Strong sensitivity of late 21st century climate to projected changes in short-lived air pollutants. Journal of Geophysical Research–Atmospheres 113(D06102).

Levy, J.I., S.M. Chemerynski, J.A. Sarnat. 2005. Ozone exposure and mortality: an empiric Bayes metaregression analysis. Epidemiology 16:458-468.

Li, Q., D.J. Jacob, I. Bey, P.I. Palmer, B.N. Duncan, B.D. Field, R.V. Martin, A.M. Fiore, R.M. Yantosca, D.D. Parrish, P.G. Simmonds, and S.J. Oltmans. 2002. Transatlantic transport of pollution and its effects on surface ozone in Europe and North America. Journal of Geophysical Research–Atmospheres 107(13).

Li, Q.Q., A. Loganath, Y.S. Chong, J. Tan, and J.P. Obbard. 2006a. Persistent organic pollutants and adverse health effects in humans. Journal of Toxicology and Environmental Health–Part A: Current Issues 69(21):1987-2005.

———. 2006b. Persistent organic pollutants and adverse health effects in humans. Journal of Toxicology and Environmental Health-Part a-Current Issues 69(21):1987-2005.

Liang, Q., L. Jaeglé, D.A. Jaffe, P. Weiss-Penzias, A. Heckman, and J.A. Snow. 2004. Long-range transport of Asian pollution to the northeast pacific: Seasonal variations and transport pathways of carbon monoxide. Journal of Geophysical Research–Atmospheres 109(D23S07):1-16.

Liang, Q., L. Jaeglé, R.C. Hudman, S. Turquety, D.J. Jacob, M.A. Avery, E.V. Browell, G.W. Sachse, D.R. Blake, W. Brune, X. Ren, R.C. Cohen, J.E. Dibb, A. Fried, H. Fuelberg, M. Porter, B.G. Heikes, G. Huey, H.B. Singh, and P.O. Wennberg. 2007. Summertime influence of Asian pollution in the free troposphere over North America. Journal of Geophysical Research 112(12):D12S11.

Lin, C.J., and S.O. Pehkonen. 1999. The chemistry of atmospheric mercury: A review. Atmospheric Environment 33(13):2067-2079.

Lin, C.J., P. Pongprueksa, S.E. Lindberg, S.O. Pehkonen, D. Byun, and C. Jang. 2006. Scientific uncertainties in atmospheric mercury models i: Model science evaluation. Atmospheric Environment 40(16):2911-2928.

Lin, C.Y., D.J. Jacob, and A.M. Fiore. 2001. Trends in exceedances of the ozone air quality standard in the continental United States, 1980-1998. Atmospheric Environment 35(19):3217-3228.

Lin, C.Y.C., D.J. Jacob, J.W. Munger, and A.M. Fiore. 2000. Increasing background ozone in surface air over the United States. Geophysical Research Letters 27(21):3465-3468.

Lin, J.-T., D.J. Donald J. Wuebbles, and X.-Z. Liang. 2008. Effects of intercontinental transport on surface ozone over the United States: Present and future assessment with a global model. Geophysical Research Letters 35(L02805, doi:10.1029/2007GL031415).

Lindberg, S., R. Bullock, R. Ebinghaus, D. Engstrom, X. Feng, W. Fitzgerald, N. Pirrone, E. Prestbo, and C. Seigneur. 2007. A synthesis of progress and uncertainties in attributing the sources of mercury in deposition. Ambio 36(1):19-32.

Lindberg, S.E., S. Brooks, C.J. Lin, K.J. Scott, M.S. Landis, R.K. Stevens, M. Goodsite, and A. Richter. 2002. Dynamic oxidation of gaseous mercury in the arctic troposphere at polar sunrise. Environmental Science and Technology 36(6):1245-1256.

Liu, B., G.J. Keeler, J.T. Dvonch, J.A. Barres, M.M. Lynam, F.J. Marsik, and J.T. Morgan. 2007. Temporal variability of mercury speciation in urban air. Atmospheric Environment 41(9):1911-1923.

Liu, J., and D.L. Mauzerall. 2005. Estimating the average time for inter-continental transport of air pollutants. Geophysical Research Letters 32:L11814, doi:10.1029/2005GL022619.

Liu, J., D.L. Mauzerall, and L.W. Horowitz. 2008. Source-receptor relationships between East Asian sulfur dioxide emissions and Northern Hemisphere sulfate concentrations. Atmospheric Chemistry and Physics 8:5537-5561.

Liu, J. D.L. Mauzerall, and L.W. Horowitz. 2009. Evaluating Inter-continental transport of fine aerosols: (2) Global Health Impacts, Atmospheric Environment 43(28):4339-4347, doi:10.1016/j.atmosenv.2009.05.032.

Loewen, M., F. Wania, F.Y. Wang, and G. Tomy. 2008. Altitudinal transect of atmospheric and aqueous fluorinated organic compounds in western Canada. Environmental Science & Technology 42(7):2374-2379.

Logan, J.A., I.A. Megretskaia, A.J. Miller, G.C. Tiao, D. Choi, L. Zhang, R.S. Stolarski, G.J. Labow, S.M. Hollandsworth, G.E. Bodeker, H. Claude, D. De Muer, J.B. Kerr, D.W. Tarasick, S.J. Oltmans, B. Johnson, F. Schmidlin, J. Staehelin, P. Viatte, and O. Uchino. 1999. Trends in the vertical distribution of ozone: A comparison of two analyses of ozonesonde data. Journal of Geophysical Research 104(D21):26373-26399.

Lohmann, R., K. Breivik, J. Dachs, and D. Muir. 2007. Global fate of pops: Current and future research directions. Environmental Pollution 150(1):150-165.

Lorenz, E.N. 1963. Deterministic nonperiodic flow. Journal of the Atmospheric Sciences 20(2):130-141.

Lucotte, M., A. Mucci, C. Hullaire-Marcel, P. Pichet, and A. Grondin. 1995. Anthropogenic mercury enrichment in remote lakes of northern Quebec (Canada). Water, Air, and Soil Pollution 80(1-4):467-476.

Luderer, G., J. Trentmann, K. Hungershöfer, M. Herzog, M. Fromm, and M.O. Andreae. 2007. Small-scale mixing processes enhancing troposphere-to-stratosphere transport by pyro-cumulonimbus storms. Atmospheric Chemistry and Physics 7(23):5945-5957.

Lyman, S., M. Gustin, E. Prestbo, P. Kilner, E. Edgerton, B. Hartsell. 2009. Testing and application of surrogate surfaces for understanding potential gaseous oxidized mercury dry deposition. Environmental Science and Technology, in press.

Lyman, S.N., M.S. Gustin, E.M. Prestbo, and F.J. Marsiks. 2007. Estimation of dry deposition of atmospheric mercury in Nevada by direct and indirect methods. Environmental Science and Technology 41(6):1970-1976.

Ma, J., and J.A. van Aardenne. 2004. Impact of different emission inventories on simulated tropospheric ozone over china: A regional chemical transport model evaluation. Atmospheric Chemistry and Physics 4:877-887.

Ma, J., S. Venkatesh, Y.-f. Li, Z. Cao, and S. Daggupaty. 2005a. Tracking toxaphene in the North American great lakes basin. 2. A strong episodic long-range transport event. Environmental Science & Technology 39(21):8132-8141.

Ma, J., S. Venkatesh, Y.-F. Li, and S. Daggupaty. 2005b. Tracking toxaphene in the North American great lakes basin. 1. Impact of toxaphene residues in United States soils. Environmental Science & Technology 39(21):8123-8131.

Ma, J.M., S. Daggupaty, T. Harner, P. Blanchard, and D. Waite. 2004. Impacts of lindane usage in the Canadian prairies on the great lakes ecosystem. 2. Modeled fluxes and loadings to the great lakes. Environmental Science & Technology 38(4):984-990.

Ma, J.M., S. Daggupaty, T. Harner, and Y.F. Li. 2003. Impacts of lindane usage in the Canadian prairies on the great lakes ecosystem. 1. Coupled atmospheric transport model and modeled concentrations in air and soil. Environmental Science & Technology 37(17):3774-3781.

Ma, J.M., and Y.F. Li. 2006. Interannual variation of persistent organic pollutants over the great lakes induced by tropical pacific sea surface temperature anomalies. Journal of Geophysical Research–Atmospheres 111(D4).

Ma, J.M., S. Venkatesh, Y.F. Li, Z.H. Cao, and S. Daggupaty. 2005c. Tracking toxaphene in the North American great lakes basin. 2. A strong episodic long-range transport event. Environmental Science & Technology 39(21):8132-8141.

Ma, J.M., S. Venkatesh, Y.F. Li, and S. Daggupaty. 2005d. Tracking toxaphene in the North American great lakes basin. 1. Impact of toxaphene residues in United States soils. Environmental Science & Technology 39(21):8123-8131.

Macdonald, R.W. 2005. Climate change, risks and contaminants: A perspective from studying the arctic. Human and Ecological Risk Assessment 11(6):1099-1104.

Macdonald, R.W., T. Harner, and J. Fyfe. 2005. Recent climate change in the arctic and its impact on contaminant pathways and interpretation of temporal trend data. Science of the Total Environment 342(1-3):5-86.

Macdonald, R.W., D. Mackay, Y.F. Li, and B. Hickie. 2003. How will global climate change affect risks from long-range transport of persistent organic pollutants? Human and Ecological Risk Assessment 9(3):643-660.

Mahowald, N.M., and C. Luo. 2003. A less dusty future? Geophysical Research Letters 30(17).

Malm, W.C., B.A. Schichtel, R.B. Ames, and K.A. Gebhart. 2002. A 10-year spatial and temporal trend of sulfate across the United States. Journal of Geophysical Research 107(D22):4627.

Mandalakis, M., H. Berresheim, and E.G. Stephanou. 2003. Direct evidence for destruction of polychlorobiphenyls by oh radicals in the subtropical troposphere. Environmental Science & Technology 37(3):542-547.

Marenco, A., H. Gouget, P. Nedelec, J.P. Pages, and F. Karcher. 1994. Evidence of a long-term increase in tropospheric ozone from Pic du Midi data series: Consequences: Positive radiative forcing. Journal of Geophysical Research–Atmospheres 99(D8):16,617-16,632.

Marenco, A., V. Thouret, P. Nédélec, H. Smit, M. Helten, D. Kley, F. Karcher, P. Simon, K. Law, J. Pyle, G. Poschmann, R. Von Wrede, C. Hume, and T. Cook. 1998. Measurement of ozone and water vapor by airbus in-service aircraft: The MOZAIC airborne program, an overview. Journal of Geophysical Research 103(D19):25631-25642.

Martin, B.D., H.E. Fuelberg, N.J. Blake, J.H. Crawford, J.A. Logan, D.R. Blake, and G.W. Sachse. 2003. Long-range transport of Asian outflow to the equatorial pacific. Journal of Geophysical Research–Atmospheres 107(8322):[printed 108(D2), 2003].

Martin, R.V. 2008. Satellite remote sensing of surface air quality. Atmospheric Environment 42(34):7823-7843.

Martin, R.V., C.E. Sioris, K. Chance, T.B. Ryerson, T.H. Bertram, P.J. Wooldridge, R.C. Cohen, J.A. Neuman, A. Swanson, and F.M. Flocke. 2006. Evaluation of space-based constraints on global nitrogen oxide emissions with regional aircraft measurements over and downwind of eastern North America. Journal of Geophysical Research 111(15): D15308.

Mason, R.P., W.F. Fitzgerald, and F.M.M. Morel. 1994. The biogeochemical cycling of elemental mercury: Anthropogenic influences. Geochimica et Cosmochimica Acta 58(15):3191-3198.

Mason, R.P., and G.R. Sheu. 2002. Role of the ocean in the global mercury cycle. Global Biogeochemical Cycles 16(4):40-1

McConnell, J.R., A.J. Aristarain, J.R. Banta, P.R. Edwards, and J.C. Simões. 2007. 20th-century doubling in dust archived in an Antarctic Peninsula ice core parallels climate change and desertification in South America. Proceedings of the National Academy of Sciences 104(14):5743-5748.

McKeen, S., G. Grell, S. Peckham, J. Wilczak, I. Djalalova, E. Hsie, G. Frost, J. Peischl, J. Schwarz, R. Spackman, A. Middlebrook, J. Holloway, J. de Gouw, C. Warneke, W. Gong, V. Bouchet, S. Gadreault, J. Racine, J. McHenry, J. McQueen, P. Lee, Y. Tang, G. Carmichael, and R. Mathur. 2009. An evaluation of real-time air quality forecasts and their urban emissions over eastern Texas during the summer of 2006 Second Texas Air Quality Study field study. Journal of Geophysical Research–Atmospheres. 114(D00F11):1-26.

McMillan, W.W., J.X. Warner, M.M. Comer, E. Maddy, A. Chu, L. Sparling, E. Eloranta, R. Hoff, G. Sachse, C. Barnet, I. Razenkov, and W. Wolf. 2008. Airs views transport from 12 to 22 July 2004 Alaskan/Canadian fires: Correlation of AIRS CO and MODIS AOD with forward trajectories and comparison of AIRS CO retrievals with DC-8 in situ measurements during INTEX-A/ICARTT. Journal of Geophysical Research 113(20):D20301.

Meehl, G.A., J.M. Arblaster, and W.D. Collins. 2008. Effects of black carbon aerosols on the Indian monsoon. Journal of Climate 21:2869-2882.

Mendoza-Dominguez, A., and A.G. Russell. 2000. Iterative inverse modeling and direct sensitivity analysis of a photochemical air quality model. Environmental Science and Technology 34:4974-4981.

Mendoza, A., M.R. Garcia, P. Vela, D.F. Lozano, and D. Allen. 2005. Trace gases and particulate matter emissions from wildfires and agricultural burning in northeastern Mexico during the 2000 fire season. Journal of the Air & Waste Management Association 55(12):1797-1808.

Merrill, J.T., and J.L. Moody. 1996. Synoptic meteorology and transport during the North Atlantic Regional Experiment (NARE) intensive: Overview. Journal of Geophysical Research–Atmospheres 101(22):28903-28921.

Meyer, T., Y.D. Lei, and F. Wania. 2006. Measuring the release of organic contaminants from melting snow under controlled conditions. Environmental Science & Technology 40(10):3320-3326.

Mickley, L.J., D.J. Jacob, B.D. Field, and D. Rind. 2004. Climate response to the increase in tropospheric ozone since preindustrial times: A comparison between ozone and equivalent CO_2 forcings. Journal of Geophysical Research 109:D05106.

Mickley, L.J., D.J. Jacob, and D. Rind. 2001. Uncertainty in preindustrial abundance of tropospheric ozone: Implications for radiative forcing calculations. Journal of Geophysical Research 106(D4):3389-3399.

Milford, J.B., and A. Pienciak. 2009. After the clean air mercury rule: Prospects for reducing mercury emissions from coal-fired power plants. Environmental Science and Technology 43(8):2669-2673.

Minguillón, M.C., M. Arhami, J.J. Schauer, and C. Sioutas. 2008. Seasonal and spatial variations of sources of fine and quasi-ultrafine particulate matter in neighborhoods near the Los Angeles-Long Beach harbor. Atmospheric Environment 42(32):7317-7328.

Moffet, R.C., Y. Desyaterik, R.J. Hopkins, A.V. Tivanski, M.K. Gilles, Y. Wang, V. Shutthanandan, L.T. Molina, R.G. Abraham, K.S. Johnson, V. Mugica, M.J. Molina, A. Laskin, and K.A. Prather. 2008. Characterization of aerosols containing zn, pb, and cl from an industrial region of mexico city. Environmental Science & Technology 42(19):7091-7097.

Moldanová, J., E. Fridell, O. Popovicheva, B. Demirdjian, V. Tishkova, A. Faccinetto, and C. Focsa. 2009. Characterisation of particulate matter and gaseous emissions from a large ship diesel engine. Atmospheric Environment 43(16):2632-2641.

Morris, G.A., S. Hersey, A.M. Thompson, S. Pawson, J.E. Nielsen, P.R. Colarco, W.W. McMillan, A. Stohl, S. Turquety, J. Warner, B.J. Johnson, T.L. Kucsera, D.E. Larko, S.J. Oltmans, and J.C. Witte. 2006. Alaskan and Canadian forest fires exacerbate ozone pollution over Houston, Texas, on 19 and 20 July 2004. Journal of Geophysical Research 111:D24S03.

Mozaffarian, D., and E.B. Rimm. 2006. Fish intake, contaminants, and human health evaluating the risks and the benefits. Journal of the American Medical Association 296(15):1885-1899.

Mudway, I.S., and F.J. Kelly. 2000. Ozone and the lung: a sensitive issue. Molecular Aspects of Medicine 21:1-48.

Mühle, J., T.J. Lueker, Y.X. Su, B.R. Miller, K.A. Prather, and R.F. Weiss. 2007. Trace gas and particulate emissions from the 2003 southern California wildfires. Journal of Geophysical Research–Atmospheres 112(3).

Muir, D.C.G., S. Backus, A.E. Derocher, R. Dietz, T.J. Evans, G.W. Gabrielsen, J. Nagy, R.J. Norstrom, C. Sonne, I. Stirling, M.K. Taylor, and R.J. Letcher. 2006. Brominated flame retardants in polar bears (ursus maritimus) from Alaska, the Canadian arctic, east Greenland, and Svalbard. Environmental Science & Technology 40(2):449-455.

Mukherjee A.B., P. Bhattacharya, A. Sarkar, R. Zevenhoven. 2008. Mercury emissions from industrial sources in India. In Mercury Fate and Transport in the Global Atmosphere: Measurements, models and policy implications, edited by N. Pirrone and R. Mason, pp. 81-112. New York: Springer.

Munthe, J., K. Kindbom, O. Kruger, G. Petersen, J. Pacyna, and Å. Iverfeldt. 2001. Examining source-receptor relationships for mercury in Scandinavia modelled and empirical evidence. Water, Air, & Soil Pollution: Focus 1(3-4):299-310.

Munthe, J., R.A. Bodaly, B.A. Branfireun, C.T. Driscoll, C.C. Gilmour, R. Harris, M. Horvat, M. Lucotte, and O. Malm. 2007. Recovery of mercury-contaminated fisheries. Ambio 36(1):33-44.

Munthe, J., E. Fjeld, M. Meili, P. Porvari, S. Rognerud, and M. Verta. 2004. Mercury in nordic freshwater fish: An assessment of spatial variability in relation to atmospheric deposition. RMZ-Mater. Geoenviron 51:1239-1241.

Murazaki, K., and P. Hess. 2006. How does climate change contribute to surface ozone change over the United States? Journal of Geophysical Research–Atmospheres 111(5).

Murphy, D.M., D.J. Cziczo, K.D. Froyd, P.K. Hudson, B.M. Matthew, A.M. Middlebrook, R.E. Peltier, A. Sullivan, D.S. Thomson, and R.J. Weber. 2006. Single-particle mass spectrometry of tropospheric aerosol particles. Journal of Geophysical Research–Atmospheres 111(D23S32).

Murphy, D. M., P. K. Hudson, D. J. Cziczo, S. Gallavardin, K. D. Froyd, M.V. Johnston, A. M. Middlebrook, M.S. Reinard, D. S. Thomson, T. Thornberry and A.S. Wexler. 2007. Distribution of lead in single atmospheric particles. Atmospheric Chemistry and Physics 7(12):3195-3210.

NARSTO. 2005. Improving emission inventories for effective air quality management across North America. NARSTO. NARSTO-05-001. Oak Ridge, TN. Available online at http://www.narsto.org/section.src?SID=8.

Neff, J.C., A.P. Ballantyne, G.L. Farmer, N.M. Mahowald, J.L. Conroy, C.C. Landry, J.T. Overpeck, T.H. Painter, C.R. Lawrence, and R.L. Reynolds. 2008. Increasing eolian dust deposition in the western United States linked to human activity. Nature Geoscience 1(3):189-195.

Nelson, P.F. 2007. Atmospheric emissions of mercury from Australian point sources. Atmospheric Environment 41:1717-1724.

Neuman, J.A., D.D. Parrish, M. Trainer, T.B. Ryerson, J.S. Holloway, J.B. Nowak, A. Swanson, F.M. Flocke, J.M. Roberts, S.S. Brown, H. Stark, R. Sommariva, A. Stohl, R.E. Peltier, R.J. Weber, A.G. Wollny, D.T. Sueper, G. Hubler, and F.C. Fehsenfeld. 2006. Reactive nitrogen transport and photochemistry in urban plumes over the North Atlantic Ocean. Journal of Geophysical Research–Atmospheres 111(23).

Newell, R.E., and M.J. Evans. 2000. Seasonal changes in pollutant transport to the north pacific: The relative importance of Asian and European sources. Geophysical Research Letters 27(16):2509-2512.

Nowak, J.B., D.D. Parrish, J.A. Neuman, J.S. Holloway, O.R. Cooper, T.B. Ryerson, D.K. Nicks, F. Flocke, J.M. Roberts, E. Atlas, J.A. de Gouw, S. Donnelly, E. Dunlea, G. Hübler, L.G. Huey, S. Schauffler, D.J. Tanner, C. Warneke, and F.C. Fehsenfeld. 2004. Gas-phase chemical characteristics of Asian emission plumes observed during itct 2k2 over the eastern North Pacific Ocean. Journal of Geophysical Research–Atmospheres 109(23):1-18.

NRC. 1998. The atmospheric sciences: Entering the twenty-first century. Washington, DC. National Academy Press.

———. 2000. Toxicological effects of methylmercury. Washington, DC: National Academy Press.

———. 2001. Global air quality an imperative for long-term observational strategies. Washington, DC: National Academy Press.

———. 2005. Radiative forcing of climate change: Expanding the concept and addressing uncertainties. Washington, DC: The National Academies Press.

———. 2008. Estimating mortality risk reduction and economic benefits from controlling ozone air pollution. Washington, DC: The National Academies Press.

———. 2009. Observing Weather and Climate from the Ground Up: A Nationwide Network of Networks. Washington, DC: The National Academies Press.

Ohara, T., H. Akimoto, J. Kurokawa, N. Horii, K. Yamaji, X. Yan, and T. Hayasaka. 2007. An Asian emission inventory of anthropogenic emission sources for the period 1980-2020. Atmospheric Chemistry and Physics 7(16):4419-4444.

Oltmans, S.J., A.S. Lefohn, J.M. Harris, I. Galbally, H.E. Scheel, G. Bodeker, E. Brunke, H. Claude, D. Tarasick, B.J. Johnson, P. Simmonds, D. Shadwick, K. Anlauf, K. Hayden, F. Schmidlin, T. Fujimoto, K. Akagi, C. Meyer, S. Nichol, J. Davies, A. Redondas, and E. Cuevas. 2006. Long-term changes in tropospheric ozone. Atmospheric Environment 40(17):3156-3173.

Oltmans, S.J., A.S. Lefohn, J.M. Harris, and D.S. Shadwick. 2008. Background ozone levels of air entering the west coast of the U.S. and assessment of longer-term changes. Atmospheric Environment 42(24):6020-6038.

Ondov, J.M., T.J. Buckley, P.K. Hopke, D. Ogulei, M.B. Parlange, W.F. Rogge, K.S. Squibb, M.V. Johnston, and A.S. Wexler. 2006. Baltimore Supersite: Highly time- and size-resolved concentrations of urban $PM_{2.5}$ and its constituents for resolution of sources and immune responses. Atmospheric Environment 40(Suppl. 2):224-237

Outridge, P.M., R.W. MacDonald, F. Wang, G.A. Stern, and A.P. Dastoor. 2008. A mass balance inventory of mercury in the Arctic Ocean. Environmental Chemistry 5(2):89-111.

Outridge, P.M., G.A. Stern, P.B. Hamilton, J.B. Percival, R. McNeely, and W.L. Lockhart. 2005. Trace metal profiles in the varved sediment of an arctic lake. Geochimica et Cosmochimica Acta 69(20):4881-4894.

Owen, R.C., O.R. Cooper, A. Stohl, and R.E. Honrath. 2006. An analysis of the mechanisms of North American pollutant transport to the central north atlantic lower free troposphere. Journal of Geophysical Research–Atmospheres 111(23).

Pacyna, E.G., J.M. Pacyna, F. Steenhuisen, and S. Wilson. 2006a. Global anthropogenic mercury emission inventory for 2000. Atmospheric Environment 40(22):4048-4063.

Pacyna, E.G., J.M. Pacyna, J. Fudala, E. Strzelecka-Jastrzab, S. Hlawiczka, and D. Panasiuk. 2006b. Mercury emissions to the atmosphere from anthropogenic sources in Europe in 2000 and their scenarios until 2020. Science of the Total Environment, 370: 147-156.

Pacyna, J.M., K. Breivik, J. Munch, and J. Fudala. 2003a. European atmospheric emissions of selected persistent organic pollutants, 1970-1995. Atmospheric Environment 37: S119-S131.

Pacyna, J.M., E.G. Pacyna, F. Steenhuisen, and S. Wilson. 2003b. Mapping 1995 global anthropogenic emissions of mercury. Atmospheric Environment 37(SUPPL. 1).

Palmer, P. I., D.J. Jacob, L.J. Mickley, D.R. Blake, G.W. Sachse, H.E. Fuelberg, and C.M. Kiley. 2003. Eastern Asian emissions of anthropogenic halocarbons deduced from aircraft concentration data. Journal of Geophysical Research 108:4753, doi:10.1029/2003JD003591.

Park, R.J., D.J. Jacob, M. Chin, and R.V. Martin. 2003. Sources of carbonaceous aerosols over the United States and implications for natural visibility. Journal of Geophysical Research–Atmospheres 108(D124355).

Park, R.J., D.J. Jacob, N. Kumar, and R.M. Yantosca. 2006. Regional visibility statistics in the United States: Natural and transboundary pollution influences, and implications for the regional haze rule. Atmospheric Environment 40(28):5405-5423.

Park, R.J., G.L. Stenchikov, K.E. Pickering, R.R. Dickerson, D.J. Allen, and S. Kondragunta. 2001. Regional air pollution and its radiative forcing: Studies with a single-column chemical and radiation transport model. Journal of Geophysical Research–Atmospheres 106(D22):28751-28770.

Parrish, D.D., J.S. Holloway, M. Trainer, P.C. Murphy, G.L. Forbes, and F.C. Fehsenfeld. 1993. Export of North America ozone pollution to the North Atlantic Ocean. Science 259(5100):1436-1439.

Parrish, D.D., Y. Kondo, O.R. Cooper, C.A. Brock, D.A. Jaffe, M. Trainer, T. Ogawa, G. Hubler, and F.C. Fehsenfeld. 2004. Intercontinental transport and chemical transformation 2002 (ITCT 2K2) and Pacific exploration of Asian continental emission (PEACE) experiments: An overview of the 2002 winter and spring intensives. Journal of Geophysical Research–Atmospheres 109(D23S01).

Parrish, D.D., D.B. Millet, and A.H. Goldstein. 2008. Increasing ozone concentrations in marine boundary layer air inflow at the west coasts of North America and Europe. Atmospheric Chemistry and Physics Discussions 8(4):13847-13901.

Parrish, D.D., M. Trainer, J.S. Holloway, J.E. Yee, M.S. Warshawsky, F.C. Fehsenfeld, G.L. Forbes, and J.L. Moody. 1998. Relationships between ozone and carbon monoxide at surface sites in the north Atlantic region. Journal of Geophysical Research–Atmospheres 103(D11):13357-13376.

Patris, N., S.S. Cliff, P.K. Quinn, M. Kasem, and M.H. Thiemens. 2007. Isotopic analysis of aerosol sulfate and nitrate during itct-2k2: Determination of different formation pathways as a function of particle size. Journal of Geophysical Research–Atmospheres 112.

Pavelin, E.G., C.E. Johnson, S. Rughooputh, and R. Toumi. 1999. Evaluation of pre-industrial surface ozone measurements made using Schonbein's method. Atmospheric Environment 33(6):919-929.

Peppler, R.A., C.P. Bahrmann, J.C. Barnard, J.R. Campbell, M.D. Cheng, R.A. Ferrare, R.N. Halthore, L.A. Heilman, D.L. Hlavka, N.S. Laulainen, C.J. Lin, J.A. Ogren, M.R. Poellot, L.A. Remer, K. Sassen, J.D. Spinhirne, M.E. Splitt, and D.D. Turner. 2000. Arm southern Great Plains site observations of the smoke pall associated with the 1998 Central American fires. Bulletin of the American Meteorological Society 81(11):2563-2591.

Perry, K.D., T.A. Cahill, R.A. Eldred, D.D. Dutcher, and T.E. Gill. 1997. Long-range transport of North African dust to the eastern United States. Journal of Geophysical Research–Atmospheres 102(10):11225-11238.

Peterson, C., and M. Gustin. 2008. Mercury in air, water and biota at the Great Salt Lake Science of the Total Environment 405:255-268.

Pétron, G., C. Granier, B. Khattatov, V. Yudin, J.F. Lamarque, L. Emmons, J. Gille, and D.P. Edwards. 2004. Monthly CO surface sources inventory based on the 2000-2001 MOPITT satellite data. Geophysical Research Letters 31(21).

Pfister, G.G., L.K. Emmons, P.G. Hess, R. Honrath, J.F. Lamarque, M. Val Martin, R.C. Owen, M.A. Avery, E.V. Browell, J.S. Holloway, P. Nedelec, R. Purvis, T.B. Ryerson, G.W. Sachse, and H. Schlager. 2006. Ozone production from the 2004 North American boreal fires. Journal of Geophysical Research 111:D24S07.

Pfister, G.G., L.K. Emmons, P.G. Hess, J.F. Lamarque, A.M. Thompson, and J.E. Yorks. 2008. Analysis of the summer 2004 ozone budget over the United States using intercontinental transport experiment ozonesonde network study (IONS) observations and model of ozone and related tracers (MOZART-4) simulations. Journal of Geophysical Research 113(23):D23306.

Piekarz, A.M., T. Primbs, J.A. Field, D.F. Barofsky, and S. Simonich. 2007. Semivolatile fluorinated organic compounds in Asian and western U.S. air masses. Environmental Science & Technology 41(24):8248-8255.

Pirrone, N. 2008. Executive summary. In Mercury fate and transport in the global atmosphere measurements, models and policy implications, edited by N. Pirrone and R. Mason, pp. xxxv-xlii. New York: Springer.

Pirrone, N., and R. Mason, eds. 2009a. Mercury Fate and Transport in the Global Atmosphere: Measurements, Models, and Policy Implications. Geneva: United Nations Environment Program.

Pirrone, N., and R. Mason, eds. 2009b. Mercury fate and transport in the global atmosphere emissions, measurements and models. New York: Springer.

Pisso, I., E. Real, K.S. Law, B. Legras, N. Bousserez, J.L. Attie, and H. Schlager. 2009. Estimation of mixing in the troposphere from Lagrangian trace gas reconstructions during long-range pollution plume transport. Journal of Geophysical Research. Accepted 2009.

Pitchford, M., W. Malm, B. Schichtel, N. Kumar, D. Lowenthal, and J. Hand. 2007. Revised algorithm for estimating light extinction from IMPROVE particle speciation data. Journal of the Air & Waste Management Association 57(11):1326-1336.

Pitts, J.N., J.A. Sweetman, B. Zielinska, R. Atkinson, A.M. Winer, and W.P. Harger. 1985. Formation of nitroarenes from the reaction of polycyclic aromatic-hydrocarbons with dinitrogen pentaoxide. Environmental Science & Technology 19(11):1115-1121.

Pochanart, P., H. Akimoto, Y. Kajii, and P. Sukasem. 2003. Carbon monoxide, regional-scale transport, and biomass burning in tropical continental Southeast Asia: Observations in rural Thailand. Journal of Geophysical Research–Atmospheres 108(17).

Pope, C.A. 2000. Invited commentary: Particulate matter-mortality exposure-response relations and threshold. American Journal of Epidemiology 152:407-412.

Pope, C.A., and D.W. Dockery. 2006. Health effects of fine particulate air pollution: Lines that connect. Journal of the Air and Waste Management Association 56:709-742.

Pope III, C.A., M. Ezzati, and D.W. Dockery. 2009. Fine-particulate air pollution and life expectancy in the United States. New England Journal of Medicine 360(4):376-386.

Poulain, A.J., J.D. Lalonde, M. Amyot, J.A. Shead, F. Raofie, and P.A. Ariya. 2004. Redox transformations of mercury in an arctic snowpack at springtime. Atmospheric Environment 38(39):6763-6774.

Pozo, K., T. Harner, S.C. Lee, F. Wania, D.C.G. Muir, and K.C. Jones. 2009. Seasonally resolved concentrations of persistent organic pollutants in the global atmosphere from the first year of the gaps study. Environmental Science & Technology 43(3):796-803.

Prather, M., M. Gauss, T. Berntsen, I. Isaksen, J. Sundet, I. Bey, G. Brasseur, F. Dentener, R. Derwent, D. Stevenson, L. Grenfell, D. Hauglustaine, L. Horowitz, D. Jacob, L. Mickley, M. Lawrence, R. von Kuhlmann, J.F. Muller, G. Pitari, H. Rogers, M. Johnson, J. Pyle, K. Law, M. van Weele, and O. Wild. 2003. Fresh air in the 21st century? Geophysical Research Letters 30(2):72-1.

Prather, M.J. 2001. Atmospheric chemistry and greenhouse gases. In Climate change 2001: The scientific basis: Contribution of Working Group I to the third assessment report of the Intergovernmental Panel on Climate Change, edited by J. T. Houghton, Y. Ding, D. J. Griggs, M. Noguer, P. J. v. d. Linden, X. Dai, K. Maskell, and C. A. Johnson, pp. 239-287. New York: Cambridge University Press.

Prestbo, E.M., and D.A. Gay. 2009. Wet deposition of mercury in the U.S. and Canada, 1996-2005: Results and analysis of the NADP mercury deposition network (mdn). Atmospheric Environment in Press, Accepted Manuscript.

Prevedouros, K., E. Brorstrom-Lunden, C.J. Halsall, K.C. Jones, R.G.M. Lee, and A.J. Sweetman. 2004. Seasonal and long-term trends in atmospheric pah concentrations: Evidence and implications. Environmental Pollution 128(1-2):17-27.

Price, H.U., D.A. Jaffe, O.R. Cooper, and P.V. Doskey. 2004. Photochemistry, ozone production, and dilution during long-range transport episodes from Eurasia to the northwest United States. Journal of Geophysical Research–Atmospheres 109:D23S13.

Primbs, T., A. Piekarz, G. Wilson, D. Schmedding, C. Higginbotham, J. Field, and S.M. Simonich. 2008a. Influence of Asian and western United States urban areas and fires on the atmospheric transport of polycyclic aromatic hydrocarbons, polychlorinated biphenyls, and fluorotelomer alcohols in the western United States. Environmental Science & Technology 42(17):6385-6391.

Primbs, T., S. Simonich, D. Schmedding, G. Wilson, D. Jaffe, A. Takami, S. Kato, S. Hatakeyama, and Y. Kajii. 2007. Atmospheric outflow of anthropogenic semivolatile organic compounds from East Asia in spring 2004. Environmental Science & Technology 41(10):3551-3558.

Primbs, T., G. Wilson, D. Schmedding, C. Higginbotham, and S.M. Simonich. 2008b. Influence of Asian and western United States agricultural areas and fires on the atmospheric transport of pesticides in the western United States. Environmental Science & Technology 42(17):6519-6525.

Prospero, J.M. 1999a. Assessing the impact of advected African dust on air quality and health in the eastern United States. Human and Ecological Risk Assessment 5(3):471-479.

———. 1999b. Long-term measurements of the transport of African mineral dust to the southeastern United States: Implications for regional air quality. Journal of Geophysical Research–Atmospheres 104(D13):15917-15927.

Prospero, J.M., P. Ginoux, O. Torres, S.E. Nicholson, and T.E. Gill. 2002. Environmental characterization of global sources of atmospheric soil dust identified with the nimbus 7 total ozone mapping spectrometer (toms) absorbing aerosol product. Reviews of Geophysics 40(1).

Prospero, J.M., and P.J. Lamb. 2003. African droughts and dust transport to the caribbean: Climate change implications. Science 302(5647):1024-1027.

Prospero, J.M., I. Olmez, and M. Ames. 2001. Al and Fe in $PM_{2.5}$ and PM_{10} suspended particles in south-central Florida: The impact of the long range transport of african mineral dust. Water Air and Soil Pollution 125(1-4):291-317.

Prospero, J.M., D.L. Savoie, and R. Arimoto. 2003. Long-term record of nss-sulfate and nitrate in aerosols on Midway Island, 1981-2000: Evidence of increased (now decreasing?) anthropogenic emissions from Asia. Journal of Geophysical Research–Atmospheres 108(D1):4019.

Qian, W., X. Tang, and L. Quan. 2004. Regional characteristics of dust storms in china. Atmospheric Environment 38(29):4895-4907.

Quinn, P.K., T.S. Bates, E. Baum, N. Doubleday, A.M. Fiore, M. Flanner, A. Fridlind, T.J. Garrett, D. Koch, S. Menon, D. Shindell, A. Stohl, and S.G. Warren. 2008. Short-lived pollutants in the arctic: Their climate impact and possible mitigation strategies. Atmospheric Chemistry and Physics 8(6):1723-1735.

Ramanathan, V., P.J. Crutzen, J. Lelieveld, A.P. Mitra, D. Althausen, J. Anderson, M.O. Andreae, W. Cantrell, G.R. Cass, C.E. Chung, A.D. Clarke, J.A. Coakley, W.D. Collins, W.C. Conant, F. Dulac, J. Heintzenberg, A.J. Heymsfield, B. Holben, S. Howell, J. Hudson, A. Jayaraman, J.T. Kiehl, T.N. Krishnamurti, D. Lubin, G. McFarquhar, T. Novakov, J.A. Ogren, I.A. Podgorny, K. Prather, K. Priestley, J.M. Prospero, P.K. Quinn, K. Rajeev, P. Rasch, S. Rupert, R. Sadourny, S.K. Satheesh, G.E. Shaw, P. Sheridan, and F.P.J. Valero. 2001. Indian Ocean experiment: An integrated analysis of the climate forcing and effects of the great indo-Asian haze. Journal of Geophysical Research–Atmospheres 106(D22):28371-28398.

Real, E., K. Law, H. Schlager, A. Roiger, H. Huntrieser, J. Methven, M. Cain, J. Holloway, J.A. Neuman, T. Ryerson, F. Flocke, J. De Gouw, E. Atlas, S. Donnelly, and D. Parrish. 2008. Lagrangian analysis of low level anthropogenic plume processing across the North Atlantic. Atmospheric Chemistry and Physics Discussions 8(2):7509-7554.

Real, E., K.S. Law, B. Weinzierl, M. Fiebig, A. Petzold, O. Wild, J. Methven, S. Arnold, A. Stohl, H. Huntrieser, A. Roiger, H. Schlager, D. Stewart, M. Avery, G. Sachse, E. Browell, R. Ferrare, and D. Blake. 2007. Processes influencing ozone levels in Alaskan forest fire plumes during long-range transport over the North Atlantic. Journal of Geophysical Research 112(10):D10S41.

Reddy, M.S., and O. Boucher. 2007. Climate impact of black carbon emitted from energy consumption in the world's regions. Geophysical Research Letters 34:L11802, doi:10.1029/2006GL028904.

Reddy, M.S., and C. Venkataraman. 2002. Inventory of aerosol and sulphur dioxide emissions from India: I–fossil fuel combustion. Atmospheric Environment 36:677-697.

Reid, J.S., E.M. Prins, D.L. Westphal, C.C. Schmidt, K.A. Richardson, S.A. Christopher, T.F. Eck, E.A. Reid, C.A. Curtis, and J.P. Hoffman. 2004. Real-time monitoring of South American smoke particle emissions and transport using a coupled remote sensing//box-model approach. Geophysical Research Letters 31:L06107, doi:10.1029/2003GL018845.

Reidmiller, D.R., A.M. Fiore, D.A. Jaffe, D. Bergmann, C. Cuvelier, F.J. Dentener, B.N. Duncan, G. Folberth, M. Gauss, S. Gong, P. Hess, J.E. Jonson, T. Keating, A. Lupu, E. Marmer, R. Park, M.G. Schultz, D.T. Shindell, S. Szopa, M.G. Vivanco, O. Wild, and A. Zuber. 2009a. The influence of foreign vs. North American emissions on surface ozone in the U.S. Atmospheric Chemistry and Physics Discussions 9(2):7927-7969.

Reidmiller, D.R., D.A. Jaffe, D. Chand, S. Strode, P. Swartzendruber, G.M. Wolfe, and J.A. Thornton. 2009b. Interannual variability of long-range transport as seen at the Mt. Bachelor observatory. Atmospheric Chemistry and Physics 9(2):557-572.

Richter, A., J.P. Burrows, H. Nüß, C. Granier, and U. Niemeier. 2005. Increase in tropospheric nitrogen dioxide over china observed from space. Nature 437(7055):129-132.

Rigby, M., R.G. Prinn, P.J. Fraser, P.G. Simmonds, R.L. Langenfelds, J. Huang, D. M. Cunnold, L.P. Steele, P.B. Krummel, R.F. Weiss, S. O'Doherty, P.K. Salameh, H.J. Wang, C.M. Harth, J. Mühle, and L.W. Porter. 2008. Renewed growth of atmospheric methane. Geophysical Research Letters 35:L22805, doi:10.1029/2008GL036037.

Robinson, A.L., N.M. Donahue, M. Shrivastava, E.A. Weitkamp, A.M. Sage, A.P. Grieshop, T.E. Lane, J.R. Pierce, and S.N. Pandis. 2007. Rethinking organic aerosol: Semivolatile emissions and photochemical aging. Science 315:1259-1262.

Robrock, K.R., P. Korytar, and L. Alvarez-Cohen. 2008. Pathways for the anaerobic microbial debromination of polybrominated diphenyl ethers. Environmental Science & Technology 42(8):2845-2852.

Rotstayn, L., and U. Lohmann. 2002. Tropical rainfall trends and the indirect aerosol effect. Journal of Climate 15(15):2103-2116.

Royal Society. 2008. Ground-level ozone in the 21st century: Future trends, impacts and policy implications. London: The Royal Society.

Saikawa, E., V. Naik, L.W. Horowitz, J. Liu, and D.L. Mauzerall. 2009. Present and potential future contributions of sulfate, black and organic carbon aerosols from China to global air quality, premature mortality and radiative forcing. Atmospheric Environment 43:2814-2822, doi:10.1016/j.atmosenv.2009.02.017, 2009.

Sakamoto, M., A. Yasutake, and H. Satoh. 2005. Methyl merury exposure in general populations of Japan, Asia and Oceania. In Dynamics of mercury pollution on regional and global scales : Atmospheric processes and human exposures around the world, edited by N. Pirrone and K. R. Mahaffey. 405-420. New York: Springer Science+Business Media, Inc.

Sakata, M., and K. Marumoto. 2005. Wet and dry deposition fluxes of mercury in Japan. Atmospheric Environment 39(17):3139-3146.

Samet, J., and D. Krewski. 2007. Health effects associated with exposure to ambient air pollution. Journal of Toxicology and Environmental Health–Part A: Current Issues 70(3-4):227-242.

Sasaki, J., S.M. Aschmann, E.S.C. Kwok, R. Atkinson, and J. Arey. 1997. Products of the gas-phase oh and no3 radical-initiated reactions of naphthalene. Environmental Science & Technology 31(11):3173-3179.

Satheesh, S.K., and V. Ramanathan. 2000. Large differences in tropical aerosol forcing at the top of the atmosphere and earth's surface. Nature 405:60-63.

Schecter, A., S. Johnson-Welch, K.C. Tung, T.R. Harris, O. Papke, and R. Rosen. 2007. Polybrominated diphenyl ether (pbde) levels in livers of U.S. human fetuses and newborns. Journal of Toxicology and Environmental Health–Part A 70(1):1-6.

Schecter, A., O. Papke, T.R. Harris, K.C. Tung, A. Musumba, J. Olson, and L. Birnbaum. 2006. Polybrominated diphenyl ether (pbde) levels in an expanded market basket survey of U.S. food and estimated pbde dietary intake by age and sex. Environmental Health Perspectives 114(10):1515-20.

Schecter, A.H., T.R. Harris, S. Brummitt, N. Shah, and O. Paepke. 2008. PBDE and HBCD brominated flame retardants in the USA, update 2008: Levels in human milk and blood, food, and environmental samples. Epidemiology 19(6):S76-S76.

Schenker, U., M. Scheringer, M. Macleod, J.W. Martin, I.T. Cousins, and K. Hungerbuhlert. 2008. Contribution of volatile precursor substances to the flux of perfluorooctanoate to the arctic. Environmental Science & Technology 42(10):3710-3716.

Scheringer, M. 2009. Long-range transport of organic chemicals in the environment. Environmental Toxicology and Chemistry 28(4):677-690.

Scheuhammer, A.M., M.W. Meyer, M.B. Sandheinrich, and M.W. Murray. 2007. Effects of environmental methylmercury on the health of wild birds, mammals, and fish. Ambio 36(1):12-18.

Schichtel, B.A., R.B. Husar, S.R. Falke, and W.E. Wilson. 2001. Haze trends over the United States, 1980-1995. Atmospheric Environment 35(30):5205-5210.

Schoeberl, M.R., J.R. Ziemke, B. Bojkov, N. Livesey, B. Duncan, S. Strahan, L. Froidevaux, S. Kulawik, P.K. Bhartia, S. Chandra, P.F. Levelt, J.C. Witte, A.M. Thompson, E. Cuevas, A. Redondas, D.W. Tarasick, J. Davies, G. Bodeker, G. Hansen, B.J. Johnson, S.J. Oltmans, H. Vömel, M. Allaart, H. Kelder, M. Newchurch, S. Godin-Beekmann, G. Ancellet, H. Claude, S.B. Andersen, E. Kyrö, M. Parrondos, M. Yela, G. Zablocki, D. Moore, H. Dier, P. von der Gathen, P. Viatte, R. Stubi, B. Calpini, P. Skrivankova, V. Dorokhov, H. de Backer, F.J. Schmidlin, G. Coetzee, M. Fujiwara, V. Thouret, F. Posny, G. Morris, J. Merrill, C.P. Leong, G. Koenig-Langlo, and E. Joseph. 2007. A trajectory-based estimate of the tropospheric ozone column using the residual method. Journal of Geophysical Research 112(24):D24S49.

Schroeder, W.H., and J. Munthe. 1998. Atmospheric mercury—an overview. Atmospheric Environment 32(5):809-822.

Schulz, M., C. Textor, S. Kinne, Y. Balkanski, S. Bauer, T. Berntsen, T. Berglen, O. Boucher, F. Dentener, S. Guibert, I.S.A. Isaksen, T. Iversen, D. Koch, A. Kirkevåg, X. Liu, V. Montanaro, G. Myhre, J.E. Penner, G. Pitari, S. Reddy, Ø. Seland, P. Stier, and T. Takemura. 2006. Radiative forcing by aerosols as derived from the aerocom present-day and pre-industrial simulations. Atmospheric Chemistry and Physics 6(12):5225-5246.

Schuster, P.F., D.P. Krabbenhoft, D.L. Naftz, L.D. Cecil, M.L. Olson, J.F. Dewild, D.D. Susong, J.R. Green, and M.L. Abbott. 2002. Atmospheric mercury deposition during the last 270 years: A glacial ice core record of natural and anthropogenic sources. Environmental Science and Technology 36(11):2303-2310.

Schwartz, J., B. Coull, F. Laden, and L. Ryan. 2008. The effect of dose and timing of dose on the association between airborne particles and survival. Environmental Health Perspectives 116:64-69.

Schwartz, J., F. Laden, and A. Zanobetti. 2002. The concentration-response relation between $PM_{2.5}$ and daily deaths. Environmental Health Perspectives 110:1025-1029.

Seigneur, C., K. Vijayaraghavan, K. Lohman, P. Karamchandani, and C. Scott. 2004. Global source attribution for mercury deposition in the United States. Environmental Science and Technology 38(2):555-569.

Selden, T.M., and D. Song. 1994. Environmental quality and development: Is there a Kuznets curve for air pollution emissions? Journal of Environmental Economics and Management 27(2):147-162.

Selin, N.E., D.J. Jacob, R.M. Yantosca, S. Strode, L. Jaeglé, and E.M. Sunderland. 2008. Global 3-d land-ocean-atmosphere model for mercury: Present-day versus preindustrial cycles and anthropogenic enrichment factors for deposition. Global Biogeochemical Cycles 22(2).

Selin, N.E., D.J. Javob, R.J. Park, R.M. Yantosca, S. Strode, L. Jaeglé, and D. Jaffe. 2007. Chemical cycling and deposition of atmospheric mercury: Global constraints from observations. Journal of Geophysical Research–Atmospheres 112(D02308).

Shindell, D. 2007. Local and remote contributions to arctic warming. Geophysical Research Letters 34(14).

Shindell, D., J.F. Lamarque, N. Unger, D. Koch, G. Faluveg, S. Bauer, and H. Teich. 2008a. Climate forcing and air quality change due to regional emissions reductions by economic sector. Atmospheric Chemistry and Physics Discussions 8(3):11609-11642.

Shindell, D.T., M. Chin, F. Dentener, R.M. Doherty, G. Faluvegi, A.M. Fiore, P. Hess, D.M. Koch, I.A. MacKenzie, M.G. Sanderson, M.G. Schultz, M. Schulz, D.S. Stevenson, H. Teich, C. Textor, O. Wild, D.J. Bergmann, I. Bey, H. Bian, C. Cuvelier, B.N. Duncan, G. Folberth, L.W. Horowitz, J. Jonson, J.W. Kaminski, E. Marmer, R. Park, K.J. Pringle, S. Schroeder, S. Szopa, T. Takemura, G. Zeng, T.J. Keating, and A. Zuber. 2008b. A multimodel assessment of pollution transport to the arctic. Atmospheric Chemistry and Physics 8(17):5353-5372.

Shindell, D.T., G. Faluvegi, A. Lacis, J. Hansen, R. Ruedy, and E. Aguilar. 2006. Role of tropospheric ozone increases in 20th-century climate change. Journal of Geophysical Research–Atmospheres 111(8):D08302.

Shoeib, M., T. Harner, and P. Vlahos. 2006. Perfluorinated chemicals in the arctic atmosphere. Environmental Science & Technology 40(24):7577-7583.

Sillman, S., F.J. Marsik, K.I. Al-Wali, G.J. Keeler, and M.S. Landis. 2007. Reactive mercury in the troposphere: Model formation and results for florida, the northeastern United States, and the atlantic ocean. Journal of Geophysical Research–Atmospheres 112(23): D23305.

Sillman, S., and P.J. Samson. 1995. Impact of temperature on oxidant photochemistry in urban, polluted rural and remote environments. Journal of Geophysical Research 100(D6):11497-11508.

Simonich, S.L., and R.A. Hites. 1995. Global distribution of persistent organochlorine compounds. Science 269(5232):1851-1854.

Simpson, J.E. 1994. Sea breeze and local winds. New York: Cambridge.

Singh, H.B., L. Salas, D. Herlth, R. Kolyer, E. Czech, M. Avery, J.H. Crawford, R.B. Pierce, G.W. Sachse, and D.R. Blake. 2007. Reactive nitrogen distribution and partitioning in the North American troposphere and lowermost stratosphere. Journal of Geophysical Research 112:D12S04, doi:10.1029/2006JD007664.

Slemr, F., E.G. Brunke, R. Ebinghaus, C. Temme, J. Munthe, I. Wängberg, W. Schroeder, A. Steffen, and T. Berg. 2003. Worldwide trend of atmospheric mercury since 1977. Geophysical Research Letters 30(10):23-1.

Slemr, F., R. Ebinghaus, C.A.M. Brenninkmeijer, M. Hermann, H.H. Kock, B.G. Martinsson, T. Schuck, D. Sprung, P. van Velthoven, A. Zahn, and H. Ziereis. 2009. Gaseous mercury distribution in the upper troposphere and lower stratosphere observed onboard the caribic passenger aircraft. Atmospheric Chemistry and Physics 9(6):1957-1969.

Smithwick, M., R.J. Norstrom, S.A. Mabury, K. Solomon, T.J. Evans, I. Stirling, M.K. Taylor, and D.C.G. Muir. 2006. Temporal trends of perfluoroalkyl contaminants in polar bears (ursus maritimus) from two locations in the North American arctic, 1972-2002. Environmental Science & Technology 40(4):1139-1143.

Sodeman, D.A., S.M. Toner, and K.A. Prather. 2005. Determination of single particle mass spectral signatures from light-duty vehicle emissions. Environmental Science and Technology 39(12):4569-4580.

Spencer, M.T., J.C. Holecek, C.E. Corrigan, V. Ramanathan, and K.A. Prather. 2008. Size-resolved chemical composition of aerosol particles during a monsoonal transition period over the Indian Ocean. Journal of Geophysical Research–Atmospheres 113(16).

Spracklen, D.V., J.A. Logan, L.J. Mickley, R.J. Park, R. Yevich, A.L. Westerling, and D.A. Jaffe. 2007. Wildfires drive interannual variability of organic carbon aerosol in the western U.S. in summer. Geophysical Research Letters 34(16).

Sprovieri, F., M. Andersson, N. Pirrone, and R. Mason. 2008. Spatial coverage and temporal trends of over-water, air-surface exchange, surface and deep sea water mercury measurements. In Mercury fate and transport in the global atmosphere measurements, models and policy implications, N. Pirrone and R. Mason, eds. 243-287. New York: Springer.

Staehelin, J., R. Kegel, and N.R.P. Harris. 1998. Trend analysis of the homogenized total ozone series of Arosa (Switzerland), 1926-1996. Journal of Geophysical Research–Atmospheres 103(D7):8389-8399.

Steffen, A., T. Douglas, M. Amyot, P. Ariya, K. Aspmo, T. Berg, J. Bottenheim, S. Brooks, F. Cobbett, A. Dastoor, A. Dommergue, R. Ebinghaus, C. Ferrari, K. Gardfeldt, M.E. Goodsite, D. Lean, A.J. Poulain, C. Scherz, H. Skov, J. Sommar, and C. Temme. 2008. A synthesis of atmospheric mercury depletion event chemistry in the atmosphere and snow. Atmospheric Chemistry and Physics 8(6):1445-1482.

Stern, D.I. 2004. The rise and fall of the environmental Kuznets curve. World Development 32(8):1419-1439.

Stevenson, D.S., F.J. Dentener, M.G. Schultz, K. Ellingsen, T.P.C. van Noije, O. Wild, G. Zeng, M. Amann, C.S. Atherton, N. Bell, D.J. Bergmann, I. Bey, T. Butler, J. Cofala, W.J. Collins, R.G. Derwent, R.M. Doherty, J. Drevet, H.J. Eskes, A.M. Fiore, M. Gauss, D.A. Hauglustaine, L.W. Horowitz, I.S.A. Isaksen, M.C. Krol, J.-F. Lamarque, M.G. Lawrence, V. Montanaro, J.-F. Müller, G. Pitari, M.J. Prather, J.A. Pyle, S. Rast, J.M. Rodriguez, M.G. Sanderson, N.H. Savage, D.T. Shindell, S.E. Strahan, K. Sudo, and S. Szopa. 2006. Multi-model ensemble simulations of present-day and near-future tropospheric ozone. Journal of Geophysical Research–Atmospheres 111:D08301, doi:10.1029/2005JD006338.

Stock, N.L., V.I. Furdui, D.C.G. Muir, and S.A. Mabury. 2007. Perfluoroalkyl contaminants in the Canadian arctic: Evidence of atmospheric transport and local contamination. Environmental Science & Technology 41(10):3529-3536.

Stohl, A. 1999. A textbook example of long-range transport: Simultaneous observation of ozone maxima of stratospheric and North American origin in the free troposphere over Europe. Journal of Geophysical Research–Atmospheres 104(D23):30445-30462.

Stohl, A., ed. 2004. Intercontinental Transport of Air Pollution. Berlin: Springer.

———. 2006. Characteristics of atmospheric transport into the arctic troposphere. Journal of Geophysical Research–Atmospheres 111(11).

Stohl, A., E. Andrews, J.F. Burkhart, C. Forster, A. Herber, S.W. Hoch, D. Kowal, C. Lunder, T. Mefford, J.A. Ogren, S. Sharma, N. Spichtinger, K. Stebel, R. Stone, J. Ström, K. Tørseth, C. Wehrli, and K.E. Yttri. 2006. Pan-arctic enhancements of light absorbing aerosol concentrations due to North American boreal forest fires during summer 2004. Journal of Geophysical Research–Atmospheres 111(22).

Stohl, A., S. Eckhardt, C. Forster, P. James, and N. Spichtinger. 2002. On the pathways and timescales of intercontinental air pollution transport. Journal of Geophysical Research–Atmospheres 107(23):4684.

Stohl, A., C. Forster, S. Eckhardt, N. Spichtinger, H. Huntrieser, J. Heland, H. Schlager, S. Wilhelm, F. Arnold, and O. Cooper. 2003. A backward modeling study of intercontinental pollution transport using aircraft measurements. Journal of Geophysical Research–Atmospheres 108(12).

Stohl, A., C. Forster, H. Huntrieser, H. Mannstein, W.W. McMillan, A. Petzold, H. Schlager, and B. Weinzierl. 2007. Aircraft measurements over Europe of an air pollution plume from southeast Asia—aerosol and chemical characterization. Atmospheric Chemistry and Physics 7(3):913-937.

Streets, D.G., Q. Zhang, and Y. Wu. 2009. Projections of global mercury emissions in 2050. Environmental Science and Technology 43(8):2983-2988.

Streets D.G., J. Hao, S. Wang, and Y. Wu. 2008. Mercury emissions from coal combustion in China. In Mercury Fate and Transport in the Global Atmosphere: Measurements, models and policy implications, edited by N. Pirrone and R. Mason, pp. 51-65. New York: Springer.

Streets, D.G., T.C. Bond, G.R. Carmichael, S.D. Fernandes, Q. Fu, D. He, Z. Klimont, S.M. Nelson, N.Y. Tsai, M.Q. Wang, J.-H. Woo, and K.F. Yarber. 2003. An inventory of gaseous and primary aerosol emissions in Asia in the year 2000. Journal of Geophysical Research 108(D21):8809, doi:10.1029/2002JD003093.

Strode, S.A., L. Jaeglé, D.A. Jaffe, P.C. Swartzendruber, N.E. Selin, C. Holmes, and R.M. Yantosca. 2008. Trans-pacific transport of mercury. Journal of Geophysical Research–Atmospheres 113(D15305).

Strode, S.A., L. Jaeglé, N.E. Selin, D.J. Jacob, R.J. Park, R.M. Yantosca, R.P. Mason, and F. Slemr. 2007. Air-sea exchange in the global mercury cycle. Global Biogeochemical Cycles 21(1).

Stull, R.B. 1988. An introduction to boundary layer meteorology, Atmospheric sciences library. Boston: Kluwer Academic Publishers.

Su, Y.S., H. Hung, P. Blanchard, G.W. Patton, R. Kallenborn, A. Konoplev, P. Fellin, H. Li, C. Geen, G. Stern, B. Rosenberg, and L.A. Barrie. 2006. Spatial and seasonal variations of hexachlorocyclohexanes (hchs) and hexachlorobenzene (hcb) in the arctic atmosphere. Environmental Science & Technology 40(21):6601-6607.

Su, Y.S., H.L. Hung, P. Blanchard, G.W. Patton, R. Kallenborn, A. Konoplev, P. Fellin, H. Li, C. Geen, G. Stern, B. Rosenberg, and L.A. Barrie. 2008. A circumpolar perspective of atmospheric organochlorine pesticides (ocps): Results from six arctic monitoring stations in 2000-2003. Atmospheric Environment 42(19):4682-4698.

Sudo, K., and H. Akimoto. 2007. Global source attribution of tropospheric ozone: Long-range transport from various source regions. Journal of Geophysical Research–Atmospheres 112(12):12302.

Sudo, K., M. Takahashi, and H. Akimoto. 2003. Future changes in stratosphere troposphere exchange and their impacts on future tropospheric ozone simulations. Geophysical Research Letters 30(24):2256, doi:10.1029/2003GL018526.

Suh, H.H., T. Bahadori, J. Vallarino, and J.D. Spengler. 2000. Criteria Air Pollutants and Toxic Air Pollutants. Environmental Health Perspectives Supplements 108(S4).

Sun, P., S. Backus, P. Blanchard, and R.A. Hites. 2006a. Temporal and spatial trends of organochlorine pesticides in great lakes precipitation. Environmental Science & Technology 40(7):2135-2141.

Sun, P., P. Blanchard, K. Brice, and R.A. Hites. 2006b. Atmospheric organochlorine pesticide concentrations near the great lakes: Temporal and spatial trends. Environmental Science & Technology 40(21):6587-6593.

Sun, P., P. Blanchard, K.A. Brice, and R.A. Hites. 2006c. Trends in polycyclic aromatic hydrocarbon concentrations in the great lakes atmosphere. Environmental Science & Technology 40(20):6221-6227.

Sun, P., Ilora, Basu, P. Blanchard, K.A. Brice, and R.A. Hites. 2007. Temporal and spatial trends of atmospheric polychlorinated biphenyl concentrations near the great lakes. Environmental Science & Technology 41(4):1131-1136.

Sunderland, E.M., D.P. Krabbenhoft, J.W. Moreau, S. Strode, and W. Landing. 2009. Mercury sources, distribution, and bioavailability in the North Pacific Ocean: Insights from data and models Global Biogeochemical Cycles in press.

Sunderland, E.M., and R.P. Mason. 2007. Human impacts on open ocean mercury concentrations. Global Biogeochemical Cycles 21(4).

Swain, E.B., D.R. Engstrom, M.E. Brigham, T.A. Henning, and P.L. Brezonik. 1992. Increasing rates of atmospheric mercury deposition in midcontinental North America. Science 257(5071):784-787.

Swartzendruber, P.C., D. Chand, D.A. Jaffe, J. Smith, D. Reidmiller, L. Gratz, J. Keeler, S. Strode, L. Jaeglé, and R. Talbot. 2008. Vertical distribution of mercury, CO, ozone, and aerosol scattering coefficient in the Pacific Northwest during the spring 2006 intex-b campaign. Journal of Geophysical Research 113:D10305.

Swartzendruber, P.C., D.A. Jaffe, E.M. Prestbo, P. Weiss-Penzias, N.E. Selin, R. Park, D.J. Jacob, S. Strode, and L. Jaeglé. 2006. Observations of reactive gaseous mercury in the free troposphere at the mount bachelor observatory. Journal of Geophysical Research 111:D24301.

Talbot, R., H. Mao, E. Scheuer, J. Dibb, and M. Avery. 2007. Total depletion of Hg° in the upper troposphere-lower stratosphere. Geophysical Research Letters 34(23).

Tanimoto, H., Y. Sawa, S. Yonemura, K. Yumimoto, H. Matsueda, I. Uno, T. Hayasaka, H. Mukai, Y. Tohjima, K. Tsuboi, and L. Zhang. 2008. Diagnosing recent CO emissions and springtime O_3 evolution in East Asia using coordinated ground-based observations of O_3 and CO during the East Asian regional experiment (EAREX) 2005 campaign. Atmospheric Chemistry and Physics Discussions 8(1):3525-3561.

Tao, L., K. Kannan, C.M. Wong, K.F. Arcaro, and J.L. Butenhoff. 2008. Perfluorinated compounds in human milk from Massachusetts, USA. Environmental Science & Technology 42(8):3096-3101.

Tegen, I. 2003. Modeling the mineral dust aerosol cycle in the climate system. Quaternary Science Reviews 22(18-19):1821-1834.

Telmer, K., and M. Veiga. 2008. Knowledge Gaps in Mercury Pollution from Gold Mining. In Mercury Fate and Transport in the Global Atmosphere: Measurements, models and policy implications, edited by N. Pirrone and R. Mason, pp. 131-172. New York: Springer.

Textor, C., M. Schulz, S. Guibert, S. Kinne, Y. Balkanski, S. Bauer, T. Berntsen, T. Berglen, O. Boucher, M. Chin, F. Dentener, T. Diehl, R. Easter, H. Feichter, D. Fillmore, S. Ghan, P. Ginoux, S. Gong, A. Grini, J. Hendricks, L. Horowitz, P. Huang, I. Isaksen, T. Iversen, S. Kloster, D. Koch, A. Kirkevåg, J.E. Kristjansson, M. Krol, A. Lauer, J.F. Lamarque, X. Liu, V. Montanaro, G. Myhre, J. Penner, G. Pitari, S. Reddy, Å. Seland, P. Stier, T. Takemura, and X. Tie. 2006. Analysis and quantification of the diversities of aerosol life cycles within aerocom. Atmospheric Chemistry and Physics 6(7):1777-1813.

Thiemens, M.H. 2006. History and applications of mass-independent isotope effects. Annual Review of Earth and Planetary Sciences 34(1):217-262.

Thompson, A.M., K.E. Pickering, R.R. Dickerson, W.G. Ellis Jr., D.J. Jacob, J.R. Scala, W.-K. Tao, D.P. McNamara, and J. Simpson. 1994. Convective transport over the central United States and its role in regional CO and ozone budgets Journal of Geophysical Research 99(D9):18,703-18,711.

Thouret, V., J.P. Cammas, B. Sauvage, G. Athier, R. Zbinden, P. Nédélec, P. Simon, and F. Karcher. 2006. Tropopause referenced ozone climatology and inter-annual variability (1994-2003) from the MOZAIC programme. Atmospheric Chemistry and Physics 6(4):1033-1051.

Toner, S.M., L.G. Shields, D.A. Sodeman, and K.A. Prather. 2008. Using mass spectral source signatures to apportion exhaust particles from gasoline and diesel powered vehicles in a freeway study using uf-atofms. Atmospheric Environment 42(3):568-581.

Toumi, R., J.D. Haigh, and K.S. Law. 1996. A tropospheric ozone-lightning climate feedback. Geophysical Research Letters 23(9):1037-1040.

Travnikov, O. 2005. Contribution of the intercontinental atmospheric transport to mercury pollution in the Northern Hemisphere. Atmospheric Environment 39(39 SPEC. ISS.):7541-7548.

Trickl, T., O.R. Cooper, H. Eisele, P. James, R. Mücke, and A. Stohl. 2003. Intercontinental transport and its influence on the ozone concentration over central Europe: Three case studies. Journal of Geophysical Research–Atmospheres 108(12).

Uno, I., K. Eguchi, K. Yumimoto, T. Takemura, A. Shimizu, M. Uematsu, Z. Liu, Z. Wang, Y. Hara, and N. Sugimoto. 2009. Asian dust transported one full circuit around the globe. Nature Geoscience 2:557-560 doi:10.1038/ngeo583.

Uno, I., Z. Wang, M. Chiba, Y.S. Chun, S.L. Gong, Y. Hara, E. Jung, S.S. Lee, M. Liu, M. Mikami, S. Music, S. Nickovic, S. Satake, Y. Shao, Z. Song, N. Sugimoto, T. Tanaka, and D.L. Westphal. 2006. Dust model intercomparison (DMIP) study over Asia: Overview. Journal of Geophysical Research–Atmospheres 111(D12213).

Usenko, S., D.H. Landers, P.G. Appleby, and S.L. Simonich. 2007. Current and historical deposition of pbdes, pesticides, pcbs, and pahs to rocky mountain national park. Environmental Science & Technology 41(21):7235-7241.

Valente, R.J., C. Shea, K. Lynn Humes, and R.L. Tanner. 2007. Atmospheric mercury in the great smoky mountains compared to regional and global levels. Atmospheric Environment 41(9):1861-1873.

van Aardenne, J.A., F.J. Dentener, J.G.J. Olivier, C.G.M.K. Goldewijk, and J. Lelieveld. 2001. A 1° × 1° resolution data set of historical anthropogenic trace gas emissions for the period 1890-1990. Global Biogeochemical Cycles 15(4):909-928.

van der A, R.J., H.J. Eskes, K.F. Boersma, T.P.C.v. Noije, M.V. Roozendael, I.D. Smedt, D.H.M.U. Peters, and E.W. Meijer. 2008. Trends, seasonal variability and dominant nox source derived from a ten year record of no2 measured from space.Journal of Geophysical Research 113(D04302, doi:10.1029/2007JD009021).

van der Werf, G.R., J.T. Randerson, L. Giglio, C.J. Collatz, P.S. Kasibhatla, and J. Arellano, A.F. 2006. Interannual variability in global biomass burning emissions from 1997 to 2004. Atmospheric Chemistry and Physics 6:3421-3441.

van Donkelaar, A., R. V. Martin, W. R. Leaitch, A.M. Macdonald, T.W. Walker, D.G. Streets, Q. Zhang, E.J. Dunlea, J.L. Jimenez, J.E. Dibb, L.G. Huey, R. Weber, and M.O. Andreae. 2008. Analysis of aircraft and satellite measurements from the Intercontinental Chemical Transport Experiment (INTEX-B) to quantify long-range transport of East Asian sulfur to Canada. Atmospheric Chemistry and Physics 8:2999-3014.

van Vuuren, D.P., J. Weyant, and F. de la Chesnaye. 2006. Multi-gas scenarios to stabilize radiative forcing. Energy Economics 28:102-120.

VanCuren, R.A. 2003. Asian aerosols in North America: Extracting the chemical composition and mass concentration of the Asian continental aerosol plume from long-term aerosol records in the western United States. Journal of Geophysical Research–Atmospheres 108(20).

VanCuren, R.A., and T.A. Cahill. 2002. Asian aerosols in North America: Frequency and concentration of fine dust. Journal of Geophysical Research–Atmospheres 107(24).

VanCuren, R.A., S.S. Cliff, K.D. Perry, and M. Jimenez-Cruz. 2005. Asian continental aerosol persistence above the marine boundary layer over the eastern north pacific: Continuous aerosol measurements from intercontinental transport and chemical transformation 2002 (itct 2k2). Journal of Geophysical Research–Atmospheres 110(9):1-19.

Vander Pol, S.S., P.R. Becker, J.R. Kucklick, R.S. Pugh, D.G. Roseneau, and K.S. Simac. 2004. Persistent organic pollutants in Alaskan murre (uria spp.) eggs: Geographical, species, and temporal comparisons. Environmental Science & Technology 38(5):1305-1312.

Varekamp, J.C., and P.R. Buseck. 1984. The speciation of mercury in hydrothermal systems, with applications to ore deposition. Geochimica et Cosmochimica Acta 48(1):177-185.

Venier, M., J. Ferrario, and R.A. Hites. 2009. Polychlorinated dibenzo-p-dioxins and dibenzo-furans in the atmosphere around the great lakes. Environmental Science & Technology 43(4):1036-1041.

Vingarzan, R. 2004. A review of surface ozone background levels and trends. Atmospheric Environment 38(21):3431-3442.

Volz, A., and D. Kley. 1988. Evaluation of the Montsouris series of ozone measurements made in the nineteenth century. Nature 332(6161):240-242.

Wallington, T.J., M.D. Hurley, J. Xia, D.J. Wuebbles, S. Sillman, A. Ito, J.E. Penner, D.A. Ellis, J. Martin, S.A. Mabury, O.J. Nielsen, and M.P.S. Andersen. 2006. Formation of c7f15cooh (pfoa) and other perfluorocarboxylic acids during the atmospheric oxidation of 8 : 2 fluorotelomer alcohol. Environmental Science & Technology 40(3):924-930.

Wang, C. 2007. Impact of direct radiative forcing of black carbon aerosols on tropical convective precipitation. Geophysical Research Letters 34(5):L05709.

Wang, H., D.J. Jacob, P. Le Sager, D.G. Streets, R.J. Park, A.B. Gilliland, and A. van Donkelaar. 2009a. Surface ozone background in the United States: Canadian and Mexican pollution influences. Atmospheric Environment 43(6):1310-1319.

Wang, K., R.E. Dickinson, and S. Liang. 2009b. Clear sky visibility has decreased over land globally from 1973 to 2007. Science 323(5920):1468-1470.

Wang, Y., D.J. Jacob, and J.A. Logan. 1998. Global simulation of tropospheric O_3-No_x-hydrocarbon chemistry 3. Origin of tropospheric ozone and effects of nonmethane hydrocarbons. Journal of Geophysical Research Atmospheres 103(D9):10757-10767.

Wängberg, I., J. Munthe, T. Berg, R. Ebinghaus, H.H. Kock, C. Temme, E. Bieber, T.G. Spain, and A. Stolk. 2007. Trends in air concentration and deposition of mercury in the coastal environment of the North Sea area. Atmospheric Environment 41(12):2612-2619.

Wania, F. 2007. A global mass balance analysis of the source of perfluorocarboxylic acids in the Arctic Ocean. Environmental Science & Technology 41(13):4529-4535.

Wania, F., and D. Mackay. 1995. A global distribution model for persistent organic-chemicals. Science of the Total Environment 161:211-232.

Wania, F., D. Mackay, and J.T. Hoff. 1999. The importance of snow scavenging of poly-chlorinated biphenyl and polycyclic aromatic hydrocarbon vapors. Environmental Science & Technology 33(1):195-197.

Warneke, C., S.A. McKeen, J.A.d. Gouw, P.D. Goldan, W.C. Kuster, J.S. Holloway, E.J. Williams, B.M. Lerner, D.D. Parrish, M. Trainer, F.C. Fehsenfeld, S. Kato, E.L. Atlas, A. Baker, and D.R. Blake. 2007. Determination of urban volatile organic compound emission ratios and comparison with an emissions database. Journal of Geophysical Research 112:D10S47, doi:10.1029/2006JD007930.

Watson, J.G., J.C. Chow, and J.E. Houck. 2001. PM2.5 chemical source profiles for vehicle exhaust, vegetative burning, geological material, and coal burning in Northwestern Colorado during 1995. Chemosphere 43:1141-1151.

Weaver, C.P., X.-Z. Liang, and J. Zhu. 2009. A Preliminary Synthesis of Modeled Climate Change Impacts on U.S. Regional Ozone Concentrations. Bulletin of the American Meteorological Society (in press).

Wedepohl, K. 1995. The composition of continental crust. Cosmochim. Acta. 59:1217-1232.

Weiss-Penzias, P., M.S. Gustin, and S.N. Lyman. 2009. Observations of speciated atmospheric mercury at three sites in Nevada: Evidence for a free tropospheric source of reactive gaseous mercury. Journal of Geophysical Research 114:D14302, doi:10.1029/2008JD011607.

Weiss-Penzias, P., D. Jaffe, P. Swartzendruber, W. Hafner, D. Chand, and E. Prestbo. 2007. Quantifying Asian and biomass burning sources of mercury using the hg/co ratio in pollution plumes observed at the mount bachelor observatory. Atmospheric Environment 41(21):4366-4379.

Weiss-Penzias, P., D.A. Jaffe, P. Swartzendruber, J.B. Dennison, D. Chand, W. Hafner, and E. Prestbo. 2006. Observations of Asian air pollution in the free troposphere at Mount Bachelor observatory during the spring of 2004. Journal of Geophysical Research–Atmospheres 111(D10304).

Weiss-Penzias, P., D.A. Jaffe, A. McClintick, E.M. Prestbo, and M. Landis. 2003. Gaseous elemental mercury in the marine boundary layer: Evidence for rapid removal in anthropogenic pollution. Environmental Science and Technology 37:3755-3763.

Wells, K.C., M. Witek, P. Flatau, S.M. Kreidenweis, and D.L. Westphal. 2007. An analysis of seasonal surface dust aerosol concentrations in the western us (2001-2004): Observations and model predictions. Atmospheric Environment 41(31):6585-6597.

West, J. J., A.F. Fiore, L.W. Horowitz, and D.L. Mauzerall. 2006. Mitigating ozone pollution with methane emission controls: global health benefits. Proceedings of the National Academy of Sciences U.S.A. 103(11):3988-3993.

White, E.M., G.J. Keeler, and M.S. Landis. 2009. Spatial variability of mercury wet deposition in eastern Ohio: Summertime meteorological case study analysis of local source influences. Environmental Science & Technology. In press.

WHO (World Health Organization). 2002. Deaths and DALYs attributable to outdoor air pollution. Available online at http://www.who.int/quantifying_ehimpacts/national/countryprofile/mapoap/en/.

———. 2003. Health risks of persistent organic pollutants from long-range transboundary air pollution. Copenhagen, Denmark: WHO Regional Office for Europe. Available online at http://www.euro.who.int.

———. 2004. Health Aspects of Air Pollution. Results From the WHO Project "Systematic Review of Health Aspects of Air Pollution in Europe." Copenhagen: WHO Regional Office for Europe.

———. 2006. Air quality guidelines: Global update 2005: Particulate matter, ozone, nitrogen dioxide, and sulfur dioxide. Copenhagen, Denmark: World Health Organization.

Wiedinmyer, C., and H. Friedli. 2007. Mercury emission estimates from fires: An initial inventory for the United States. Environmental Science and Technology 41(23):8092-8098.

Wild, O., and H. Akimoto. 2001. Intercontinental transport of ozone and its precursors in a 3-D global CTM. Journal of Geophysical Research 106:27,729-27,744, doi:10.1029/2000JD000123.

Wild, O., and M.J. Prather. 2000. Excitation of the primary tropospheric chemical mode in a global three-dimensional model. Journal of Geophysical Research 105(D20):24647-24660.

Wild, O., M.J. Prather, and H. Akimoto. 2001. Indirect long-term global radiative cooling from NO_x emissions. Geophysical Research Letters 28(9):1719-1722.

Wild, O., P. Pochanart, and H. Akimoto. 2004a. Trans-Eurasian transport of ozone and its precursors. Journal of Geophysical Research–Atmospheres 109(11):D11302.

Wild, O., M.J. Prather, H. Akimoto, J.K. Sundet, I.S.A. Isaksen, J.H. Crawford, D.D. Davis, M.A. Avery, Y. Kondo, G.W. Sachse, and S.T. Sandholm. 2004b. Chemical transport model ozone simulations for spring 2001 over the western pacific: Regional ozone production and its global impacts. Journal of Geophysical Research–Atmospheres 109(15).

Wild, O., J.K. Sundet, M.J. Prather, I.S.A. Isaksen, H. Akimoto, E.V. Browell, and S.J. Oltmans. 2003. Chemical transport model ozone simulations for spring 2001 over the Western Pacific: Comparisons with trace-p lidar, ozonesondes, and total ozone mapping spectrometer columns. Journal of Geophysical Research–Atmospheres 108(21):8826, doi:10.1029/2002JD003283.

WMO. 2004. The changing atmosphere : An integrated global chemistry observation theme for the igos partnership. Report of the Integrated Global Atmospheric Chemistry Observation Theme Team. September. World Meteorological Organization. Report GAW No. 159 (WMO TD No. 1235). Noordwijk, the Netherlands.

Wong, F., H.A. Alegria, T.F. Bidleman, V. Alvarado, F. Angeles, A.A. Galarza, E.R. Bandala, I.D. Hinojosa, I.G. Estrada, G.G. Reyes, G. Gold-Bouchot, J.V.M. Zamora, J. Murguia-Gonzalez, and E.R. Espinoza. 2009. Passive air sampling of organochlorine pesticides in mexico. Environmental Science & Technology 43(3):704-710.

Wotawa, G., P.C. Novelli, M. Trainer, and C. Granier. 2001. Inter-annual variability of summertime co concentrations in the Northern Hemisphere explained by boreal forest fires in North America and Russia. Geophysical Research Letters 28(24):4575-4578.

Wotawa, G., and M. Trainer. 2000. The influence of Canadian forest fires on pollutant concentrations in the United States. Science 288(5464):324-328.

Wu, S., L.J. Mickley, D.J. Jacob, D. Rind, and D.G. Streets. 2008a. Effects of 2000-2050 changes in climate and emissions on global tropospheric ozone and the policy-relevant background surface ozone in the United States. Journal of Geophysical Research 113:D18312.

Wu, S., L.J. Mickley, E.M. Leibensperger, D.J. Jacob, D. Rind, and D.G. Streets. 2008b. Effects of 2000-2050 global change on ozone air quality in the United States. Journal of Geophysical Research 113:D06302.

Wu, Y., S. Wang, D.G. Streets, J. Hao, M. Chan, and J. Jiang. 2006. Trends in anthropogenic mercury emissions in china from 1995 to 2003. Environmental Science and Technology 40(17):5312-5318.

Wuebbles, D.J., and K. Hayhoe. 2002. Atmospheric methane and global change. Earth-Science Reviews 57(3-4):177-210.

Xin, M., M. Gustin, and D. Johnson. 2007. Laboratory investigation of the potential for re-emission of atmospherically derived hg from soils. Environmental Science and Technology 41(14):4946-4951.

Xu, X., W. Lin, T. Wang, P. Yan, J. Tang, Z. Meng, and Y. Wang. 2008. Long-term trend of surface ozone at a regional background station in eastern China 1991-2006: Enhanced variability. Atmospheric Chemistry and Physics Discussions 8(1):215-243.

Yang, H., N.L. Rose, R.W. Battarbee, and J.F. Boyle. 2002. Mercury and lead budgets for lochnagar, a scottish mountain lake and its catchment. Environmental Science and Technology 36(7):1383-1388.

Yienger, J.J., M. Galanter, T.A. Holloway, M.J. Phadnis, S.K. Guttikunda, G.R. Carmichael, W.J. Moxim, and H. Levy II. 2000. The episodic nature of air pollution transport from Asia to North America. Journal of Geophysical Research 105(D22):26931-26945.

Yu, H., L. A. Remer, M. Chin, H. Bian, R. G. Kleidman, and T. Diehl. 2008. A satellite-based assessment of transpacific transport of pollution aerosol. Journal of Geophysical Research 113:D14S12, doi:10.1029/2007JD009349.

Zeng, X., S.L.M. Simonich, K.R. Robrock, P. Korytar, L. Alvarez-Cohen, and D.F. Barofsky. 2008. Development and validation of a congener-specific photodegradation model for polybrominated diphenyl ethers. Environmental Toxicology and Chemistry 27(12):2427-2435.

Zhang, L., D.J. Jacob, K.F. Boersma, D.A. Jaffe, J.R. Olson, K.W. Bowman, J.R. Worden, A.M. Thompson, M.A. Avery, R.C. Cohen, J.E. Dibb, F.M. Flocke, H.E. Fuelberg, L.G. Huey, W.W. McMillan, H.B. Singh, and A.J. Weinheimer. 2008a. Transpacific transport of ozone pollution and the effect of recent Asian emission increases on air quality in North America: An integrated analysis using satellite, aircraft, ozonesonde, and surface observations. Atmospheric Chemistry and Physics Discussions 8(2):8143-8191.

Zhang, L.S., J.M. Ma, S. Venkatesh, Y.F. Li, and P. Cheung. 2008b. Modeling evidence of episodic intercontinental long-range transport of lindane. Environmental Science & Technology 42(23):8791-8797.

Zhang, Q., D.G. Streets, G.R. Carmichael, K. He, H. Huo, A. Kannari, Z. Klimont, I. Park, S. Reddy, J.S. Fu, D. Chen, L. Duan, Y. Lei, L. Wang, and Z. Yao. 2009a. Asian emissions in 2006 for the NASA INTEX-B mission. Atmospheric Chemistry and Physics Discussions 9:4081-4139.

Zhang, Y., X. Wang, H. Chen, X. Yang, J. Chen, and J.O. Allen. 2009b. Source apportionment of lead-containing aerosol particles in shanghai using single particle mass spectrometry. Chemosphere 74(4):501-507.

Zhang, Y.X., and S. Tao. 2009. Global atmospheric emission inventory of polycyclic aromatic hydrocarbons (PAHs) for 2004. Atmospheric Environment 43(4):812-819.

Zhao, T.L., S.L. Gong, X.Y. Zhang, and D.A. Jaffe. 2008. Asian dust storm influence on North American ambient PM levels: Observational evidence and controlling factors. Atmospheric Chemistry and Physics 8(10):2717-2728.

Zhu, J.X., Y. Hirai, S. Sakai, and M.H. Zheng. 2008. Potential source and emission analysis of polychlorinated dibenzo-p-dioxins and polychlorinated dibenzofurans in china. Chemosphere 73(1):S72-S77.

Ziemke, J.R., S. Chandra, B.N. Duncan, L. Froidevaux, P.K. Bhartia, P.F. Levelt, and J.W. Waters. 2006. Tropospheric ozone determined from Aura OMI and MLS: Evaluation of measurements and comparison with the global modeling initiative's chemical transport model. Journal of Geophysical Research 111(19):D19303.

Appendix A

Committee Sponsors, Statement of Task, and Schedule

The committee's work was sponsored by four U.S. Federal agencies: Environmental Protection Agency, National Oceanic and Atmospheric Administration, National Aeronautics and Space Administration, and National Science Foundation.

STATEMENT OF TASK

This study will summarize the state of knowledge regarding the international flows of air pollutants into and out of the United States and across its various regions, on continental and intercontinental scales. It will also consider the impact of these flows on the achievement of environmental policy objectives related to air quality or pollutant deposition in the United States and abroad and impacts on regional and global climate change. The pollutants to be considered include ozone and its precursors, fine particles and their precursors, mercury, and persistent organic pollutants. The committee will address the following core questions:

1. How does international transport of air pollutants (including ozone, aerosols, mercury, and POPs) into the United States on continental and intercontinental scales affect air quality, pollutant deposition, and radiative forcing?

—With respect to ozone and aerosols, how are exceedances of the National Ambient Air Quality Standards (NAAQS) for ozone and fine particles affected by changes in emissions in other countries?

—With respect to mercury and POPs, how are pollutant deposition and U.S. population exposure affected by changes in emissions in other countries?

—What is the level of confidence in these estimates?

2. How are foreign emissions sources expected to change in the future and how might these changes affect achievement of environmental policy objectives in the United States related to air quality, pollutant deposition, and radiative forcing?

3. How does international transport of air pollutants out of the United States affect air quality, pollutant deposition, radiative forcing, and the achievement of related environmental policy objectives in other parts of the world?

4. What additional research, observations, analysis, and information management efforts, are needed to better understand and quantify the impacts and implications of the international transport of air pollutants?

Although the committee is encouraged to provide quantitative information to the extent possible, qualitative analysis and discussions may prove appropriate for some topics, especially where the degree of confidence is uncertain. Local scale air pollution issues within shared international transboundary airsheds (such as El Paso-Juarez and Detroit-Windsor) will not be addressed.

ACTIVITIES AND SCHEDULE

The Committee met five times over the course of 18 months from June 2008 to April 2009. These meetings encompassed discussions of the Statement of Task and its context with the study sponsors and other government stakeholders, reviews of the existing literature, presentations from a variety of scientific experts, and evaluation and synthesis of this information into a final consensus report. This report was submitted for peer review in June 2009 and approved for publication in August 2009.

Appendix B

Technical Discussion of Atmospheric Transport Mechanisms

Once an air pollutant is released into the atmosphere, chemical, microphysical, and meteorological factors determine how it is distributed. The location of air pollution sources with respect to local, regional, and global air circulation patterns influences how efficiently pollutants are transported and dispersed. The winds transport air both horizontally and vertically. Vertical transport is important when considering long-range pollutant transport because pollutants distributed to higher altitudes usually encounter stronger winds that provide rapid transport to distant locations. Atmospheric stability, controlled by how temperature varies with height, determines whether vertical transport will be slow or rapid. After emission, pollutants may undergo chemical transformation, be subjected to depletion processes such as particle scavenging and dry or wet deposition, or mix into the atmosphere to become a component of the background concentration. This appendix provides a general description of the atmosphere and a synopsis of air circulation and weather patterns that influence the distribution of air pollutants.

VERTICAL STRUCTURE

Pollutant transport occurs in the lowest two layers of the atmosphere— the troposphere and stratosphere. Most weather phenomena that affect pollutant transport occur in the troposphere, which extends from the surface to ~ 18 km in the tropics and ~ 8 km near the poles (Figure B.1). The tropopause is the zone of transition between the troposphere and stratosphere. The height of the tropopause does not uniformly decrease in the poleward

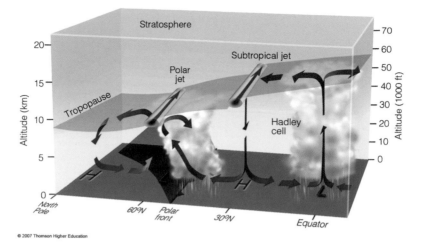

© 2007 Thomson Higher Education

FIGURE B.1 Cross-section from the equator to the North Pole showing three circulation cells, the tropopause, and the polar and subtropical jet streams. SOURCE: Ahrens, 2007.

direction. Instead, there are two climatologically occurring breaks in each hemisphere, one containing the subtropical jet stream (near 30° N), and the other containing the polar jet stream (near 45° N). These jet streams are not stagnant in location, but shift with season, moving closer to the equator in winter and more poleward during summer.

The jet streams are important distributors of air pollutants for two reasons. They create major areas of air exchange between the troposphere and stratosphere. And the strong winds of the jet stream can rapidly transport pollutants. For example, if one assumes an average wind speed of 35 m s^{-1} ($\sim$ 70 kt) at 40° latitude, an eastward-moving air parcel will circumnavigate the globe in only 10 days. The stratosphere, which extends to $\sim$ 50 km, has regions of strong winds, but virtually no turbulent mixing except for occasional overshooting thunderstorms, certain types of lightning, and occasional thin clouds.

The atmosphere's vertical temperature profile plays the dominant role in controlling whether and how quickly an air pollutant will be dispersed upward from its point of emission. The change of temperature with height or lapse rate is used to quantify vertical temperature profiles. The average midlatitude tropospheric temperature lapse rate is 6.5°C km^{-1}, with actual values constantly changing in both time and three-dimensional space. Large lapse rates (like those near the surface on a sunny day) are associated with atmospheric instability, which promotes turbulence. Conversely, small lapse rates near the surface (as would occur on a cold, windless night) denote stability that suppresses vertical motion. Layers containing temperature inversions (a negative lapse rate) are very stable, greatly inhibiting vertical transport and promoting the accumulation of pollutants. Inversions in the troposphere can occur when the surface is colder than the overlying air and in subsiding air, which occurs in regions of high pressure. The stratosphere is a permanently stable region, with a near zero lapse rate between 11 and 20 km and increasingly negative (stable) rates above. As a result of this stability pollutants injected into the stratosphere tend to remain there for much longer periods than in the troposphere.

An important characteristic of free tropospheric air movement is that air parcels experiencing no exchange of heat energy conserve their potential temperature and thus move along surfaces of constant potential temperature (isentropic surfaces). Exceptions are regions of cloud cover where radiative processes and water vapor phase changes can be major sources or sinks of heat. In addition, air parcels in the surface boundary layer undergo temperature changes due to exchanges of radiation with Earth's surface. Isentropic surfaces slope upward toward the north, with isentropic values increasing vertically (Figure B.2). As a result poleward moving air conserving its potential temperature tends to ascend, while equatorward-moving air tends to sink. This concept has applications for pollution transport into the Arctic (Stohl et al., 2006; Law and Stohl, 2007). Specifically, pollution-laden parcels beginning at low altitudes and heading north that conserve their potential temperature will ascend to the middle troposphere. Conversely, for low-level parcels to remain near the surface during northward excursions, they either must be very cold initially or undergo considerable loss of heat due to passing over ice-covered surfaces, especially during the long polar winter seasons. Northern Eurasia is sufficiently cold that its pollutants can be transported quasi-horizontally to the Arctic, making it a major source of Arctic pollution during winter.

GLOBAL CIRCULATION FEATURES

Global circulation patterns are driven by the nonuniform distribution of incoming solar energy, with the greatest energy being received near the equator. The general circulation can be considered the multiyear seasonal average of the daily winds. The smaller, shorter-lived circulations described later are removed, leaving the largest, longest-lasting wind patterns. The flow in the tropical troposphere ($\pm 0°$-$30°$) is dominated by the Hadley Circulation Cell (Figure B.3) which contains rising air along the Intertropical Convergence Zone (ITCZ). This ascent produces a band of enhanced clouds and precipitation. Subsiding air and relatively clear skies occur at the poleward boundary of the Hadley Cell. A component of this sinking air moves southward to replace the air that has ascended up and

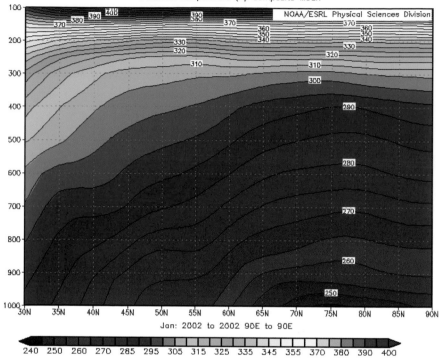

FIGURE B.2 Cross-section of mean potential temperature (K) from 30° N to the North Pole. Note that the isentropes slope upward toward the pole. Air parcels move along isentropic surfaces when no heat energy is added or subtracted.

away from the equator. Thus, the meridional flow is equatorward at low levels and poleward above. The middle latitudes (30°-60°) are dominated by transient cyclones (low-pressure areas) and anticyclones (high-pressure areas), especially during the winter. In the long term mean the region is characterized by sinking air near 30° and rising air at its northern boundary (~ 60°), corresponding to the location of the polar front. This region sometimes is denoted the Ferrell Cell (not depicted in Figure B.3). The polar troposphere (60°-90°) is dominated by rising air near 60° and sinking air over the poles, sometimes denoted the Polar Cell.

The simplified view of the global circulation described above becomes more complex when the effects of continents and oceans are included. Global sea-level pressure patterns and surface winds for January and July are shown in Figure B.4. Focusing on the Northern Hemisphere, the Janu-

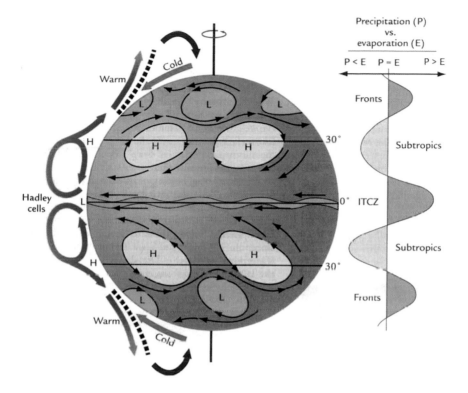

FIGURE B.3 An idealized representation of Earth's general circulation.
SOURCE: Ahrens, 2007.

ary pattern (Figure B.4a) is dominated by high-pressure air masses with clockwise circulating winds over the relatively cold Eurasian and North American continents. The Bermuda and Pacific high-pressure regions are evident but weak. Conversely, the Icelandic and Aleutian Lows represent the average of transient synoptic-scale low-pressure systems that form near the east coasts of Asia and North America and then move eastward, reaching maximum intensity near the location of lowest pressure in the figure. Air circulates counterclockwise around these lows. One should note the ITCZ that extends around the globe just south of the equator; it represents the confluence of the northeasterly trade winds (Northern Hemisphere) with the southeasterly trades (Southern Hemisphere) and is an important area of interhemispheric transport.

Global circulations during the Northern Hemisphere summer (Figure B.4b) are quite different from those during winter. Low pressure, not high pressure, now dominates the continents, producing the seasonal wind

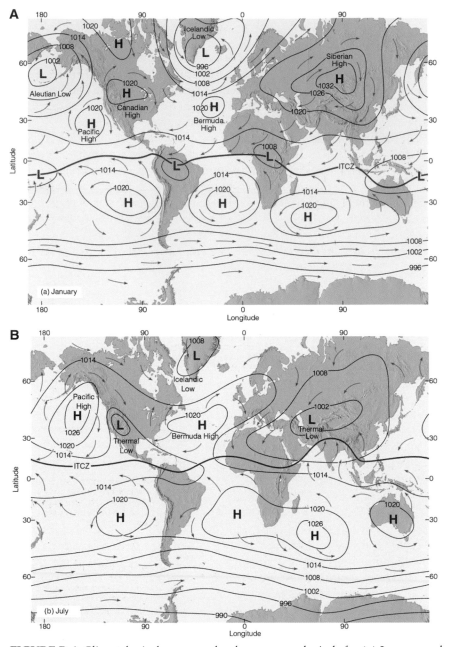

FIGURE B.4 Climatological mean sea-level pressure and winds for (a) January and (b) July. The Intertropical Convergence Zone (ITCZ) is shown by the red line near the equator.
SOURCE: Ahrens, 2007.

reversal called the monsoon. In Asia, for example, there is offshore flow during the winter but onshore flow during summer. The quasi-permanent Bermuda and Pacific high-pressure regions are larger and better defined during summer than winter. Their southern extents produce the northeasterly trade winds, which combined with their Southern Hemisphere counterpart, produce the ITCZ that now is located north of the equator. The Icelandic and Aleutian storm tracks are poorly defined because their constituent synoptic-scale transient lows are much weaker during summer.

Global flow patterns in the middle and upper troposphere (Figure B.5) are simpler than those near the surface. Prominent features are easterly flow in the deep tropics, clockwise flow around the semipermanent high-pressure regions in the subtropics, and circumpolar cyclonic flow. The prevailing westerlies, which contain north to south undulations cover a major portion of the Northern Hemisphere. The westerlies are stronger during winter than summer.

The following points summarize the role of global circulations in producing long-range transport.

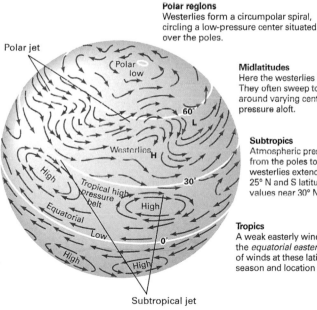

Polar regions
Westerlies form a circumpolar spiral, circling a low-pressure center situated over the poles.

Midlatitudes
Here the westerlies become very strong. They often sweep to the north or south around varying centers of high and low pressure aloft.

Subtropics
Atmospheric pressure continues to rise from the poles toward the Equator, hence westerlies extend equatorward to about 25° N and S latitude, with maximum values near 30° N and 30° S.

Tropics
A weak easterly wind pattern prevails, called the *equatorial easterlies*. However, the directic of winds at these latitudes can shift with the season and location on the globe.

FIGURE B.5 Climatological flow patterns in the middle troposphere.
SOURCE: Anderson and Strahler, 2008. Reproduced with permission of John Wiley and Sons Inc.

• Winds in the middle-latitude troposphere are mostly from the west (zonal flow), causing most intercontinental transport to be from west to east.

• The north-south (meridional) component of the wind in the middle and upper troposphere usually is much weaker than the zonal component. The two components can have similar magnitudes near the surface.

• Wind speeds generally are stronger during winter than summer, causing more rapid transport during the winter months. The jet streams in the upper troposphere are regions of strongest winds.

• Wind speeds in the troposphere generally increase with altitude. Thus, the vertical motion experienced by air parcels is vitally important since pollutants that are transported from near the surface to higher altitudes usually will be horizontally transported the most rapidly. Areas of rising air tend to be smaller and shorter lived than areas of subsidence, which generally cover larger areas and persist longer.

SYNOPTIC SYSTEMS

Synoptic circulation features have sizes of ~ 1,000-2,000 km and lifetimes of several days to a week. Transient middle-latitude cyclones (lows) and anticyclones (highs) are prime examples of these circulations. Anticyclones generally are regions of tranquil weather with sinking air that leads to relatively cloud-free skies and stable conditions that suppress mixing and tend to trap pollutants. Their light winds also reduce horizontal transport. Anticyclones with little forward motion allow these stagnating conditions to persist over days or even weeks.

Low-pressure areas are important regions of strong horizontal and vertical pollution transport. Locations of cyclone initiation (cyclogenesis) and their subsequent storm tracks are important in determining the routes of long-range pollution transport. Once a cyclone begins to form it is "steered" by upper tropospheric flow patterns, generally toward the east. Important areas of cyclogenesis are located over eastern Asia and the western Pacific Ocean, as well as the east coast of North America. Cyclones forming in these areas are important mechanisms for transporting pollutants from the east coasts of both Asia and North America (Merrill and Moody, 1996; Cooper et al., 2002a,b; Stohl et al., 2002). Another preferred region of cyclogenesis is downwind of major mountain ranges such as the Rocky Mountains or the Alps. It is noteworthy that Europe and western Asia are not major regions of cyclone formation or transit.

The instantaneous flow around cyclones in the northern hemisphere is counterclockwise. However, if one considers three dimensional trajectories with respect to a moving cyclone, three specific pathways (or airstreams) often are identified—the warm and cold conveyor belts and the dry intrusion (Figure B.6) (Browning and Monk, 1982; Browning and Roberts, 1994; Bader, 1995; Carlson, 1998). The warm conveyor belt (WCB) is a major transporter of pollutants (Stohl et al., 2002; Eckhardt et al., 2004). It begins near the surface in advance of the cyclone's cold front (i.e., its warm sector). If the cyclone forms sufficiently offshore, relatively clean maritime air is transported by the WCB. If the low forms closer to land, surface-based pollutants from the heavily industrialized regions of eastern Asia and eastern North America are transported by WCBs. The pollution-laden air rises slowly at first but more quickly as it approaches the cyclone's warm front. The thunderstorms that sometimes are embedded within the WCB can produce localized regions of much more rapid ascent (Kiley and Fuelberg, 2006). When the air has ascended to the middle troposphere, it begins to move eastward and become part of the background westerly flow. By the end of the conveyor the air typically has reached the altitude of the tropopause (~ 9 km). The transport time from the boundary layer near the east coast of the United States to the European free troposphere typically is three to four days (Stohl et al., 2002; Eckhardt et al., 2004), but can be as short as two days if the jet stream is particularly strong (Stohl et al., 2003). Due to the greater distance for transpacific transport, an extra day or two may be required to move pollution from East Asia to North America, again depending on the strength of the jet stream (Cooper et al., 2004). In some cases a second cyclone may be involved.

As its name implies the cold conveyor belt is located completely within the cold sector of the cyclone (Figure B.6). The low-level air flows toward the west along the north (cold side) of the surface warm frontal position. During part of this route, the WCB is overhead. As cold air approaches the center of the cyclone the air begins to ascend into the middle troposphere while making a clockwise loop, eventually reversing direction and combining with the WCB in the upper troposphere. The role of the cold conveyor belt in transporting pollutants aloft has received relatively little attention.

The dry air intrusion (DI) of a middle-latitude cyclone originates in the upper troposphere and lower stratosphere (Figure B.6). It is located on the poleward side of the cyclone and descends into the middle to lower troposphere. The DI is characterized by subsidence and often by regions of much lower tropopause height (tropopause folds) that are related to the jet stream aloft. Thus, the DI can transport upper tropospheric or stratospheric air into the middle or lower troposphere. Some authors have described a cold, dry post-cold-frontal airstream in the middle to lower troposphere beneath the DI and behind the surface cold front (Cooper et al., 2001).

It is noteworthy that air masses also can be transported long distances without being lifted (i.e., the air and its pollutants remain in the lower troposphere). This generally occurs in the absence of transient synoptic systems that would contain mechanisms for ascent (e.g., the WCB). Arctic

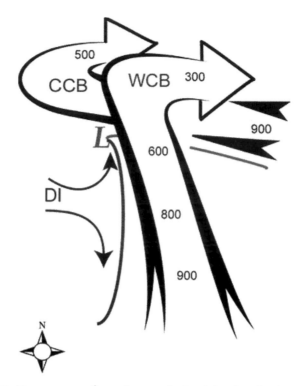

FIGURE B.6 Airstream configuration as depicted in the classic cyclone model (adapted from Carlson, 1998). Airstreams are the warm conveyor belt (WCB), cold conveyor belt (CCB), and dry intrusion (DI). Numbers indicate the approximate pressure altitudes (hPa) of the airstreams. The surface low-pressure center is indicated with an "L". The lines extending south and east of the low-pressure center indicate the surface cold front and warm front, respectively.
SOURCE: Kiley and Fuelberg, 2006.

haze (Barrie, 1986) has been attributed to this low-level transport (Klonecki et al., 2003; Stohl, 2006; Law and Stohl, 2007). It also has been observed downwind of North America over the North Atlantic Ocean (Neuman et al., 2006), the Azores (Owen et al., 2006), and Europe (Li et al., 2002; Guerova et al., 2006). Similar phenomena have been observed over the North Pacific Ocean (Liang et al., 2004; Holzer et al., 2005) and the Indian Ocean during the winter monsoon (Ramanathan et al., 2001).

MESOSCALE SYSTEMS

Mesoscale weather systems have typical sizes of a few hundred kilometers and lifetimes ranging from a few hours to a day. Important examples associated with pollutant transport are thunderstorms, land and sea circulations, and mountain and valley breezes. These circulations either can be superimposed on the larger scale transient systems or they can occur alone.

Thunderstorms occur frequently over many parts of the world, ranging from isolated cells to organized clusters called mesoscale convective systems (MCSs). The bases of thunderstorms typically are ~ 1.5 km above the surface, while the tops of nonsevere isolated cells or disorganized clusters extend to near the local tropopause. Updrafts and downdrafts generally are less than 10 m s⁻¹. The structure and life cycle of a typical nonsevere thunderstorm is shown in Figure B.7. These storms can rapidly move boundary layer pollutants to the upper troposphere where they can be transported great distances by the stronger horizontal winds aloft (Dickerson et al., 1987; Lelieveld and Crutzen, 1994). Conversely, the downdrafts that occur during the mature and dissipating stages of a storm transport upper tropospheric air to the surface. Nonsevere storms can be associated with cyclones and frontal systems or be embedded within homogeneous synoptic air masses. A prime example is Florida and surrounding states, which experience almost daily thunderstorms during the warm season.

Examples of severe convection include supercells, multicell complexes, and squall lines. Doswell (2001) presents an excellent summary of severe convective storms. These storms have three important characteristics that

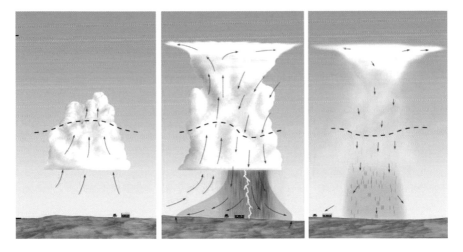

FIGURE B.7 Life cycle of a typical nonsevere thunderstorm. Updrafts are shown as red arrows and downdrafts by blue arrows. The storm initially (left panel) contains only updrafts but contains only downdrafts during dissipation (right panel). SOURCE: Ahrens, 2007.

relate to atmospheric transport. First, their updrafts are much stronger than their nonsevere counterparts, often reaching 40 m s^{-1}, allowing the storms to overshoot the tropopause and extend several kilometers into the stratosphere. These strong updrafts can transport boundary layer air to the upper troposphere or lower stratosphere on the order of minutes, compared with hours or days for synoptic systems. This is an important consideration for short-lived chemical species. Second, the structure of severe storms differs from that of nonsevere storms. For example, the cross-section through a mature squall in Figure B.8 reveals a rear inflow jet that transports mid-level air downward and toward the surface. Supercell storms and multicell systems (not shown) have somewhat different structures. Severe storms generally have a long lifetime, often lasting 12 h or more. Thus, the storms can move long distances during their lifetimes and produce strong vertical transport over a large area for an extended time.

A newly discovered type of convection is associated with wildfires. These pyroconvection events can transport large quantities of aerosols and gases into the upper troposphere and lower stratosphere (Damoah et al., 2006; Fromm et al., 2000, 2005; Jost et al., 2004; Luderer et al., 2007).

In summary, deep convection is a very efficient transporter of boundary layer air to the free troposphere (Dickerson et al., 1987; Park et al., 2001). Cotton et al. (1995) estimated an annual flux of 4.95×10^{19} kg of boundary layer air by cloud systems (including extratropical cyclones), which represents a venting of the entire boundary layer about 90 times a year. Calculations for the central United States suggest that nearly 50 percent of boundary layer CO is transported to the free troposphere by deep convection during summer (Thompson et al., 1994), while a typical middle latitude squall line was found to transport of 9.9×10^3 tons of CO out of

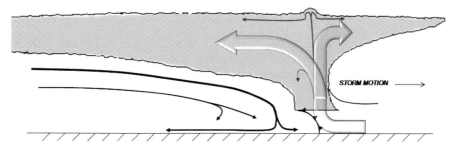

FIGURE B.8 Cross section perpendicular to a squall line (adapted from Houze et al., 1989). Large hollow arrows identify the ascending front to rear inflow (left) and core updraft transporting air to the cloud top and forward anvil (right). Black arrows represent the rear inflow jet supporting the cold pool generation directly below the core updraft.

the boundary layer over an 8 h simulation period, 3.89×10^4 t past 500 hPa, and 2.88×10^4 t of CO above 300 hPa (Halland et al., 2009).

Sea and land breezes are important sources of mesoscale transport in all three dimensions. They are examples of mesoscale diurnally varying thermal circulations that form due to temperature contrasts between the land and adjacent ocean (Simpson, 1994). During the warm part of the day, the land surface is warmer than the ocean, producing onshore surface flow (the sea breeze). The extent of inland penetration is greatly influenced by the direction of the prevailing larger scale wind. The flow above the sea breeze is reversed (offshore winds) to complete the circulation cell, with the depth of the complete circulation usually confined to the lowest 3 km of the atmosphere. The leading edge of the advancing low level sea breeze is a region of strong ascent that often produces thunderstorms in humid regions of the world. At night the temperature gradient reverses, causing offshore flow near the surface (the land breeze) and onshore wind aloft. The land breeze usually is much weaker than its daytime counterpart. Sea and land breezes can transport coastal emissions offshore during the day and onshore during the night.

Mountain and valley circulations also are diurnally varying mesoscale thermal circulations. In this case the horizontal temperature gradient is due to altitude differences. During the day, the mountains act as an elevated heat source, causing air and its pollutants to rise up the side of the sloping terrain. Under summertime fair weather conditions three times the volume of the valley can be lofted into the free troposphere each day (Henne et al., 2004). If conditions are favorable, the ascent can lead to thunderstorm development along the mountain tops. At higher altitudes away from the mountains the air sinks into the surrounding valleys. At night the horizontal temperature gradient reverses, producing downslope flow into the nearby valley that is assisted by gravity. This can lead to an accumulation of pollutants in the valley.

Without mountain and valley circulations mountain ranges could block the horizontal transport of pollutants. If the daytime upslope flow is sufficiently strong, the polluted air can rise up and over the mountains and be transported away from its source by the stronger winds aloft. Mexico City is a prime example of where terrain-induced circulations strongly affect pollution concentrations (Fast et al., 2007; Lei et al., 2007; De Foy et al., 2008).

MICROSCALE MOTIONS

Turbulence is the prime example of microscale circulations. Turbulence is important in pollution transport because it can thoroughly mix the air and its pollutants. A well-mixed layer is characterized by vertically uniform concentrations of pollutants, water vapor, and potential temperature. The depth of the mixed layer is denoted the planetary boundary layer (PBL).

There are two major categories of surface-based turbulence. Mechanically-induced turbulence occurs when the prevailing horizontal wind is disrupted by a rough surface. Thermally-induced turbulence occurs when the temperature lapse rate is large, producing a relatively unstable surface layer and causing the unevenly heated surface to produce pockets of ascent and descent. Since over land both the prevailing wind speed and temperature lapse rate typically are strongest during the day, mixing generally is stronger during the day than the night. Therefore, the height of the mixed layer also varies diurnally (Stull, 1988). There is much less diurnal variation over water.

Boundary layer turbulence appears to be the major source of vertical transport in parts of Asia. Dickerson et al. (2007) found that the warm-sector PBL air ahead of a cold front was highly polluted while in the free troposphere, concentrations of trace gases and aerosols were less but still well above background. They concluded that dry convection appears to dominate vertical transport, with warm conveyor belts first coming into play as the cyclonic systems move off the coast.

As the PBL collapses with the onset of evening, pollutants that were transported aloft by turbulence remain, forming a residual layer (Stull, 1988) that is decoupled from the surface and thereby experiences stronger wind speeds than air within the PBL (Angevine et al. 1996). Most turbulence in the free troposphere (above the PBL) is produced by an optimum combination of temperature lapse rate and the degree to which the prevailing winds vary with height (wind shear). Although the forcing mechanisms differ, the effect is the same—turbulence in the free atmosphere thoroughly mixes the air.

TRACKING AIR PARCELS—TRAJECTORY APPROACHES

It often is important to determine the source of air pollution at a specific location or where pollution from a given source will be located in the future. The basic concept is simple if one has an accurate four-dimensional (x,y,z,t) representation of temperature and wind. One simply uses the data to advect air parcels backward or forward over increments of time. Accurately applying this concept is very difficult because of the myriad types and scales of processes that affect transport.

The isentropic and kinematic methods are the most widely used procedures for calculating trajectories. The isentropic method assumes that an air parcel conserves its potential temperature during the computational period. Thus, the parcel is advected on its sloping isentropic surface by the horizontal winds. The vertical component of the wind is not needed in these calculations since parcels are assumed to change altitude because of the slope of the isentropic surface. The isentropic assumption generally is very good in the stratosphere for periods of a week or longer since there are no surface radiative processes and few clouds. Nonetheless, radiative processes increasingly violate the isentropic assumption over time. The isentropic assumption is violated much more quickly in the troposphere, where it is usually not the preferred methodology for calculating trajectories.

The kinematic method utilizes the three dimensional wind components at an initial location to advect air parcels over an interval of time. Once at the new location and time the wind components at that location and time are used to advect the parcel. The process continues for the desired time period. As with isentropic trajectories the wind data are from numerical meteorological models whose grid spacing typically varies from ~ 50 to 150 km for global data, down to a few km for regional models. The models typically contain approximately 50 levels in the vertical, often with closer spacing near the surface and near tropopause level. An example of 10-day backward trajectories is shown in Figure B.9.

Particle dispersion models are an advanced version of the trajectory concept. They have been widely used in relatively recent transport studies. A well-known example is the FLEXPART model (Stohl et al., 2005). Dispersion models require three-dimensional wind components from either coupled or off-line meteorological models, and may contain modules that seek to incorporate the effects of convection and other sub-grid-scale motions that are not adequately represented by the input meteorological

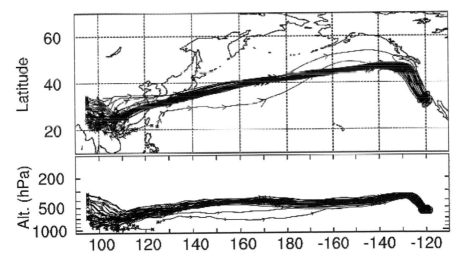

FIGURE B.9 Cluster of 10-day backward trajectories arriving just offshore of Southern California and initiating over Southeast Asia. The upper panel is a plan view; the lower panel is longitude versus altitude (hPa). The trajectories arrive at 0000 UTC March 12, 1999.
SOURCE: Adapted from Martin et al. (2003).

data (e.g., Forster et al., 2007). In addition, each particle that is released at a source can be assigned a mass that is related to the rate of emission. Thus, maps showing future concentrations of a species can be produced. An example of a FLEXPART run is given in Figure B.10.

Trajectories and particle dispersion models require a perfect depiction of the atmosphere at every time step to be totally accurate. This is a corollary to the famous statement that every time a butterfly flaps its wings, its motion will ultimately affect the weather (Lorenz, 1963). Unfortunately, however, vast areas of Earth have sparse or even no surface-based observations. Although satellite remote sensing reduces the problem, some synoptic systems still are inadequately resolved, with smaller systems being diagnosed even less accurately. As an example, individual thunderstorms and their updrafts and downdrafts will not be resolved by a global atmospheric model having a horizontal resolution of 50 km, or even a regional model at a resolution of 10 km. Instead, the models will utilize parameterization schemes to diagnose the composite effects of the storms at the scale of the model. Parameterization schemes also are used to simulate the effects of boundary layer processes, radiative effects, and other processes. Inadequate numerical techniques to compute the trajectories are another factor limiting the accuracy of trajectories.

As a result of our inability to completely describe all atmospheric motions, trajectories (and weather forecasts) deteriorate with time. The exact rate of deterioration is very difficult to quantify since it depends on

2 July 2004, 9-12 UTC

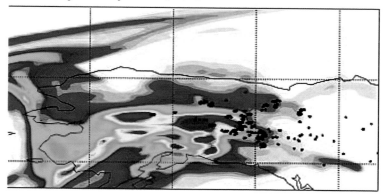

FIGURE B.10 Total column of CO tracer from the forward FLEXPART simulation shown for July 2, 2004, 0900–1200 UTC. Black dots show MODIS fire detections on the respective day.
SOURCE: Adapted from Stohl et al. (2006).

the types of weather phenomena that are occurring and how well they are detected and parameterized.

INTERCONTINENTAL POLLUTION TRANSPORT

Figure 1.2 in Chapter 1 depicts the major pathways of pollution transport in the Northern Hemisphere, considering first the transpacific transport from Asia toward North America. Modeling studies indicate that the transport occurs year round (Liang et al., 2004), but is strongest during spring when three to five Asian plumes affect the boundary layer of the west coast of the United States between February and May (Yienger et al., 2000). This is due to the frequency and structure of the eastward-moving middle-latitude cyclones and the exact paths they take. Strong Asian plumes have been observed by aircraft over the eastern North Pacific Ocean (Heald et al., 2003; Nowak et al., 2004) and the west coast of the United States (Jaffe et al., 1999, 2003a; Jaeglé et al., 2003; Cooper et al., 2004). Most of the plumes were associated with lifting of East Asian pollutants by the WCBs of middle-latitude cyclones. As noted previously, there is considerable stratospheric-tropospheric exchange and general subsidence to the rear of the cyclones. As the cyclones decay along the west coast of North America, the plumes dissipate and become part of the hemispheric pollution background. Some Asian plumes have remained sufficiently intact that they have been detected over Europe (Stohl et al., 2007).

Most of the North American export toward the east also is associated with middle-latitude cyclones and their associated WCBs (Figure 1.1 in Chapter 1). Evidence of North American pollution has been observed in the European free troposphere (Stohl, 1999; Stohl et al., 2003; Trickl et al., 2003) and at high-altitude surface sites in the Alps (Huntrieser et al., 2005). Weak effects of North American pollution have been detected at Mace Head, Ireland (Derwent et al., 2007). Forest fires over Alaska and Canada have produced greater enhancements of low-level concentrations at Mace Head (Forster et al., 2001).

There is no major cyclonic storm track between Europe and Asia (Figure B.4), and few studies have examined the transport of European pollution to Asia (Newell and Evans, 2000; Pochanart et al., 2003; Duncan and Bey, 2004; Wild et al., 2004a). Newell and Evans (2000) estimated that on an annual basis, only 24% of the air parcels arriving over Central Asia had passed over Europe, with 4% originating in the European PBL. European pollution also has been detected over eastern Siberia (Pochanart et al., 2003), Japan (Wild et al., 2004), and North Africa (Lelieveld et al., 2002; Stohl et al., 2002). Instead, European emissions are exported at relatively low altitudes and strongly affect the Arctic (Stohl et al., 2002; Duncan and Bey, 2004).

SUMMARY

The sections above indicate that many meteorological phenomena on a variety of spatial and temporal scales transport surface pollutants out of the boundary layer and into the free troposphere, including thunderstorms, turbulence, sea breezes, and the warm conveyor belts of cyclones.[1] Donnel et al. (2001) found that advection was the most important mechanism for transporting tracer to the free troposphere; and the addition of upright convection and turbulent mixing increased the amount by up to 24 percent, with convection transporting the tracer to heights of 5 km. They concluded that the convection and turbulent mixing were not linearly additive processes, emphasizing the importance of representing all such processes in meteorological modeling studies.

More generally, the long-range intercontinental transport of pollutants can be considered a two-step process. First, the pollutants must be transported vertically out of the boundary layer where winds are relatively light and into the free troposphere where winds are stronger, especially near the jet stream. Once in the free troposphere the pollutants are transported quasi-horizontally by larger wind systems such as the prevailing westerlies. The strength of the winds determines how rapidly the transport will occur, and there can be considerable mixing with stratospheric air above the troposphere.

Many middle-latitude low-pressure areas form near the highly industrialized east coast of Asia. Their WCBs can carry the pollutants aloft where they are transported quasi-horizontally toward the west coast of North America. If convection occurs near the low-pressure area, the upward transport occurs much more rapidly. The low pressure development is episodic, occurring approximately every four days during the winter and spring but less often during the warm season. Therefore, the pollution tends to traverse the North Pacific in elongated bursts or plumes before becoming part of the background concentration at even greater distances from their Asian source.

The heavily populated east coast of the United States also is a region of enhanced low pressure development. Similar to that described for eastern Asia, the WCBs, and possibly convection associated with the developing lows, vertically transport the pollutants out of the polluted boundary layer where they are carried eastward toward Europe. Transport from Europe to Asia occurs mainly in the lower troposphere because Europe is not a major region of low pressure development. However, when deep convec-

---

[1] In contrast, there is relatively little quantitative information comparing the relative roles of gravitational settling, scavenging by precipitation, and other processes that transport pollutants back down to the surface. This is an area that requires additional research.

tion occurs, low level pollution can be quickly transported aloft into the westerlies.

The transient low pressure systems described above are middle latitude phenomena. Transport from the Sahara to the far southeastern United States occurs at lower latitudes and is due to quasi-permanent subtropical high pressure located over the Atlantic Ocean (Bermuda and Azores Highs). The clockwise flow around these systems produces easterly winds that provide the westward transport.

The Arctic lower troposphere is isolated from the rest of the atmosphere by its very cold air, i.e., the Arctic front. However, the front is not zonally symmetric, and can extend to 40°N over Eurasia during January. Thus, northern Eurasia is the major source of Arctic pollution during winter. Air from further south can be transported to the Arctic, but only in the middle and upper troposphere. During summer, the transport is from the North Atlantic Ocean, across the high Arctic, and toward the North Pacific.

Appendix C

Observational Platforms Used for Long-Range Pollution Transport Studies

	Ozone	Ozone Precursors[a]	PM	PM (chemical constituents)	Hg[b]	POPs	Key Contributions
Commercial Aircraft							
MOZAIC	Y	Y					Routine cross-Atlantic measurements on several European Airbus aircraft in flight corridor plus profiles of O_3, CO, H_2O (1994-present) and NO, NOy in addition (2001-present).
CARIBIC	Y	Y	Y		Y		Multiple species measured in container on an in-service aircraft as part of European project, 1997 to 2002 and 2004 to present.
NOXAR	Y	Y					Routine measurement of O_3, NO, NO_2 on Swiss aircraft over North Atlantic in UT in 1995/1996.
Airborne Campaigns [*North American export*]							
NARE I & II	Y	Y	Y	Y			Large international field campaigns involving multiple platforms examining export of North American pollution over North Atlantic, summer 1993 and 1997.
SONEX	Y	Y					NASA-led campaign focused on impact of aircraft emissions on UT(LS), autumn 1997.
POLINAT	Y	Y					German DLR-Falcon campaign on impact of aircraft emissions over North Atlantic, autumn 1997.

					Description
ACTO, MAXOX, EXPORT	Y	Y			European multi-aircraft campaigns examining import of NA pollution into European free troposphere and export of EU pollution, spring and summer 1999-2000.
ICARTT	Y	Y	Y	Y	Large international field campaign involving multiple platforms focused on detection and quantification of North America to Europe long-range transport. First attempt to conduct Lagrangian experiment, by making multiple sampling of same plume across Atlantic. Included NASA INTEX-NA, European ITOP missions, summer 2004.
YAK-AEROSIB	Y	Y			French-Russian flights over Siberia, export of European pollution, transport of Siberian FF plumes to NA, spring and summer 2006-2008.
ABLE 3A/3B	Y	Y			GTE missions investigating chemical composition over tundra and Arctic regions, 1988, 1990.
TOPSE	Y	Y	Y		Characterization of Arctic Haze breakdown, origins of springtime ozone in midlatitude and polar regions, winter and spring 2000

continued

	Ozone	Ozone Precursors[a]	PM	PM (chemical constituents)	Hg[b]	POPs	Key Contributions
POLARCAT	Y	Y	Y	Y	Y		Large international field on export of NA (and other) pollution to the Arctic, radiative and climate impacts – contribution to IPY; impact of AN and FF pollution. Included NASA ARCTAS, DLR-GRACE, NOAA-ICEALOT, NOAA-ARCPAC, DOE-ISDAC, POLARCAT-France missions, spring and summer 2008.
Airborne Campaigns [*focus on North American import*]							
PEM-West B	Y	Y					NASA experiment to quantify outflow of pollutants from Asia during spring 1994.
TRACE-P/ACE-Asia	Y	Y	Y	Y			Large international field experiment involving numerous platforms and groups to examine Asian pollutant formation, outflow and transport across the Pacific, spring 2001.
ITCT-2K2	Y	Y	Y	Y			NOAA-led campaign focusing on import of pollutants into southwest USA, summer 2002.
INTEX-B	Y	Y	Y	Y	Y		NASA campaign focusing on pollution import into North America (also DLR-Falcon over Germany), spring 2006.
PHOBEA-I, II	Y	Y	Y		Y	Y	University of Washington experiment (1997-1999 and 2001-2003) to detect and quantify flux of Asian pollution to North America.

Satellite Instruments

Satellite Instruments			UV absorbing (dust, smoke)	
METEOR-3, NIMBUS-7, EP/TOMS	Y		Y	Multiyear dataset for the study of interannual variability of tropospheric O_3 and AI.
ERS-2/GOME	Y	Y	Y	Multiyear dataset for the study of interannual variability and inverse modeling of emissions, especially NO_2.
ENVISAT/SCIAMACHY	Y	Y	Y	Multiyear dataset for the study of interannual variability and inverse modeling of emissions, especially NO_2.
TERRA/MOPITT		Y		Multiyear CO dataset for the study of interannual variability and inverse modeling of emissions.
TERRA, AQUA/MODIS	Y		Y	Multiyear AOD dataset for the study of interannual variability; daily coverage is possible for tracking aerosol plumes as they develop; inverse modeling of emissions.
TERRA/MISR			Y	Aerosol type and plume height information.
AQUA/AIRS	Y	Y	Y	Daily coverage is possible for CO plume tracking; upper troposphere O_3.
(SCISAT/ACE)	Y	Y	Y	Observations of transport and chemical processing in the upper troposphere.
AURA/TES	Y	Y	Y	Coincident measurements of O_3 and precursors for process studies.

continued

	Ozone	Ozone Precursors[a]	PM	PM (chemical constituents)	Hg[b]	POPs	Key Contributions
AURA/OMI	Y	Y	Y	Dust, smoke, sulfates			Good coverage of O_3, precursors and AI; possibility of tracking plumes as they develop; inverse modeling of emissions.
AURA/MLS	Y	Y					Good sensitivity to upper tropospheric CO transport; provides stratospheric O_3 column for use in tropospheric O_3 residual retrievals (especially with OMI).
METOP-A/IASI	Y	Y					Daily coverage of O_3 and precursors is possible for plume tracking; first of 3 missions that will provide data through 2020.
METOP-A/GOME2	Y	Y					First of 3 missions that will provide data through 2020.
(CLOUDSAT/CALIPSO)			Y				High vertical resolution information on aerosol distribution and type for suborbital plume transects .
Ground-based networks							
NOAA/ESRL Atmospheric Baseline Observatories	Y	Y	Y				5 sites (3 in NH) focused on GHG distribution. Some use of this data for LRT studies.
NOAA/ESRL Cooperative Air Sampling Network		Y					Network collects weekly flask samples from approximately 100 sites around the globe. Major focus is on GHG distribution, but data provide some utility for understanding LRT.

| WMO/GAW Global Observatories | Y | Y | There are approximately 22 GAW stations, which includes the ESRL sites and other national programs. GAW observatories are similar in design to ESRL Atmospheric Baseline Observatories but provide broader spatial coverage. Gas phase observations are made regularly at most sites. The GAW program recommends an extensive set of aerosol measurements, including AOD, light scattering, size distribution. However, only a small fraction of the sites are currently reporting aerosol data to the GAW data center. |
| CASTNET Clean Air Status and Trends Network | Y | | U.S. network of approximately 100 sites to obtain background observations of surface ozone and some additional parameters relevant to understanding trends and ecosystem impacts of air pollutants. A small fraction of the sites measure CO and NOy. Many sites are colocated with an IMPROVE aerosol sampler. |

continued

	Ozone	Ozone Precursors[a]	PM	PM (chemical constituents)	Hg[b]	POPs	Key Contributions
IMPROVE Interagency Monitoring of Protected Visual Environments			Y	Y			U.S. network of approximately 200 sites to obtain regional background aerosol mass loadings and chemical composition of fine mode aerosol (D < 2.5 μm). IMPROVE data has been used in numerous studies to document LRT. Many sites are colocated with a CASTNET measurements of O_3.
CAMNET Canadian Atmospheric Mercury Network					Y (TGM and wet dep)		Canadian Atmospheric Mercury Network (CAMNET) measures total gaseous mercury at 11 sites across Canada. Wet deposition of mercury is also measured at some of these sites.
MDN/AMI Mercury Deposition Network, Atmospheric Mercury Initiative					Y (wet dep)		MDN collects weekly Hg in wet deposition at approximately 85 sites in the United States. AMI is a new inititive which will collect speciated Hg data at multiple sites for quantification of dry deposition.
EMEP	Y	Y	Y	Y	Y	Y	European program to measure gas and aerosol compounds in support of the Convention on Long-Range Transboundary Air Pollution. Measurements are made at several hundred sites across Europe.

EANET (Acid Deposition Monitoring Network in East Asia)	Y				Y	Integration of national networks across Asia for acid deposition, ozone, and aerosols.
Other Research sites						
Mt. Bachelor Observatory	Y	Y	Y	Y	Y	Mountain top (2.7 km) research station on the west coast of the United States (Oregon). Operated by the University of Washington and Oregon State University.

[a] Ozone precursors include CO, CH_4, volatile organic compounds, and nitrogen oxides. A [Y] in this box indicates measurement of any one or more of these species.

[b] While the chemical form of Hg is important, here we do not distinguish between measurements of total, elemental or Hg+2.

Appendix D

Committee Membership

Dr. Charles E. Kolb Jr. (*Chair*) is the president and chief executive officer of Aerodyne Research Inc. (ARI), a position he has held since 1984. Since 1970 ARI has provided research and development services requiring expertise in the physical and engineering sciences to commercial and government clients working to solve national and international environmental problems. These include a wide range of topics such as global and regional environmental quality and the development of clean and efficient energy and new propulsion technologies. Dr. Kolb has received numerous professional honors and has served in a broad range of professional and Academy-related positions. He has been elected a fellow of the American Geophysical Union, the American Physical Society, the Optical Society of America, and the American Association for the Advancement of Science. He is currently chair of the Advisory Council for the Department of Civil and Environmental Engineering at Princeton University and has served as chair of the Committee for Environmental Improvement of the American Chemical Society (2000-2008). He has contributed to a variety of National Academies studies and is currently serving as a member of the National Research Council's Board on Chemical Science and Technology. Dr. Kolb holds an S.B. in chemistry from the Massachusetts Institute of Technology and an M.S. and Ph.D. in physical chemistry from Princeton University. His research interests include atmospheric, combustion, and materials chemistry as well as physics and chemistry of aircraft and rocket exhaust plumes. In addition to over 250 reports, nonrefereed symposia papers, patents, book reviews, and policy papers, Dr. Kolb has published over 190 archival journal articles and book chapters.

Dr. Tami Bond is an associate professor of civil and environmental engineering and an affiliate professor in atmospheric sciences at the University of Illinois, Urbana-Champaign. Dr. Bond's research addresses the aerosol chemistry, physics, and optics that govern the environmental impacts of combustion effluents. Her research includes development of past, present, and future global emission inventories, global simulations of aerosol transport and fate, and laboratory and field measurements of particle emission rates and properties. Dr. Bond is a member of American Geophysical Union and American Association for Aerosol Research, and an editor at *Aerosol Science and Technology*. She holds a B.S. in mechanical engineering from the University of Washington, an M.S. in mechanical engineering (University of California, Berkeley) and an interdisciplinary Ph.D. in atmospheric sciences, civil engineering, and mechanical engineering (University of Washington, 2000). She held a NOAA Climate and Global Change Postdoctoral Fellowship before joining the University of Illinois in 2003.

Dr. Mae S. Gustin is an associate professor in the Department of Natural Resources and Environmental Sciences at the University of Nevada in Reno. Her primary research interest is the study of the fate and transport of contaminants in the environment. Dr. Gustin's current work focuses primarily on sources and sinks for atmospheric mercury and the pathways by which atmospheric mercury is input to ecosystems. Specific research topics include quantifying the contribution of natural sources of mercury to the atmosphere; understanding soil-plant-air mercury exchange processes; investigating fugitive mercury emissions from active gold mines; characterizing the role of plants in biogeochemical cycling of mercury; development of surrogate surfaces for measuring atmospheric mercury dry deposition as well as passive samplers for characterization of air-mercury speciation; measurement of air-mercury speciation and dry deposition at locations in Nevada, Utah, and the southeastern United States; and characterization of mercury concentrations, water quality, and sources of mercury to select reservoirs in Nevada. Other environmental contaminants of research include arsenic, trifluoroacetic acid and organophosphate pesticides. Her regular teaching responsibilities include the following courses: Environmental Pollution (sophomore level), NRES 467/667 Regional and Global Issues in Environmental Science (senior capstone), and NRES 765 Biogeochemical Cycles (graduate student class) Dr. Gustin received her Ph.D. from the University of Arizona in economic geology and geochemistry in 1988.

Dr. Gregory R. Carmichael, professor of chemical and biological engineering at the University of Iowa, is a leader in the development of emissions inventories for natural and pollutant substances and of chemical transport models at scales ranging from local to global. He has worked extensively on issues of long-range transport of acidic and photochemical pollutants from Asia and on the impact of Asian development on the environment. He is an active instructor and adviser, having supervised 29 M.S. and 24 Ph.D. students. Dr. Carmichael received his Ph.D. in chemical engineering from the University of Kentucky in 1979. He has served as department chair and is director of the Center for Global and Regional Environmental Research. He is presently chair of the Scientific Advisory Committee of the WMO Urban Environment Research Program and serves on the steering committee of the Commission on Atmospheric Chemistry and Global Pollution. He has been a member and chair of the American Meteorological Society's Committee on Atmospheric Chemistry and on numerous other committees and boards. Dr. Carmichael has over 220 refereed journal publications and serves on a number of editorial boards.

Dr. Kristie L. Ebi currently serves as executive director of the Intergovernmental Panel on Climate Change Technical Support Unit for Working Group II. She previously served as senior managing scientist in Exponent's Health Sciences Center for Epidemiology, Biostatistics, and Computational Biology. Dr. Ebi specializes in research on the potential impacts of global environmental change, and on the design of adaptation response options to reduce negative impacts. She designs, conducts, and interprets scientific investigations, including analyses and evaluation of data and literature. She has worked on a range of issues related to the potential health impacts of global climate change, including impacts associated with vector borne diseases, heat waves, extreme events (flooding), food-borne diseases, and air pollution. She conducts research on the potential health impacts of residential and occupational exposures to magnetic fields. Examples of the studies on which she has worked range from the possible role of magnetic field exposure in childhood leukemia to occupational magnetic field exposures in garment workers. Dr. Ebi is a lead author for the Human Health chapter in the Fourth Assessment Report of the Intergovernmental Panel on Climate Change. She was lead author in Working Group II (Response Options) of the Millennium Ecosystem Assessment (Chapters on Human Health and on Uncertainties in Assessing the Effectiveness of Response Options Regarding Ecosystem Services). She was a lead author of the Health Sector Analysis Team of the U.S. National Assessment of the Potential Consequences of Climate Variability and Change, and was a contributing author to the Human Health Chapter of the Third Assessment Report of the Intergovernmental Panel on Climate Change.

Dr. David P. Edwards is a senior scientist and group leader in the Atmospheric Chemistry Division at the National Center for Atmospheric Research (NCAR). He is also a NASA investigator and the NCAR Project Leader for the MOPITT instrument on the NASA Terra satellite, with management responsibilities for data processing, algorithm enhancement, data validation exercises, and coordinating science investigations. His research concentrates on the scientific utilization of tropospheric remote sensing data with emphasis on the cross-scale combination of measurements from multiple satellite platforms, aircraft, and ground stations. He is a coauthor of a white paper submitted to the NRC Decadal Survey charged with determining the priorities for the next round of missions for Earth Science and Applications from Space. He received his Ph.D. in dense plasma physics theory from the University of Birmingham in 1987.

Dr. Henry E. Fuelberg is a professor of meteorology at Florida State University. His research is in the areas of synoptic and mesometeorology. Dr. Fuelberg's current NASA research projects examine the long-range transport of pollutants from Asia, Mexico, and Alaskan wildfires using data from past and upcoming NASA airborne field projects. Dr. Fuelberg also has close ties with the National Weather Service, studying ways to improve thunderstorm and lightning forecasts in the southeastern United States improved forecasts of Florida lightning also are being sponsored by Florida Power & Light Corp. He is preparing a high-resolution (4 * 4 km, hourly) historical precipitation database (1996-current) for the Florida Department of Environmental Protection that will be used in precipitation and hydrologic studies. He received his Ph.D. in meteorology in 1976 from Texas A&M University.

Dr. Jiming Hao is a professor in the Department of Environmental Engineering at Tsinghua University and dean of the Research Institute of Environmental Science and Engineering. He received his Ph.D. in 1984 in environmental engineering from the University of Cincinnati. He serves as vice editor-in-chief for the *Journal of the Air & Waste Management Association,* and is a member of the Chinese Academy of Engineering. Dr. Hao has published nearly 200 academic papers and a number of monographs on a wide variety of issues related to air pollution control engineering.

Dr. Daniel J. Jacob is the Vasco McCoy Family Professor of Atmospheric Chemistry and Environmental Engineering at Harvard University. He was formerly the Gordon McKay Professor of Atmospheric Chemistry and Environmental Engineering (1994-2004). Dr. Jacob's research focuses on understanding the composition of the atmosphere, its perturbation by human activity, and the implications for human welfare and climate. He has served on the NASA Earth Systems Science and Applications Advisory Committee and has been lead or co-lead scientist on several NASA aircraft missions. He is also the lead scientist for the GEOS-CHEM chemical transport model used by a large number of research groups in North America and Europe. Dr. Jacob is the recipient of the NASA Distinguished Public Service Medal (2003) and the American Geophysical Union James B. Macelwane Medal (1994). He was chair of the NRC Committee on Radiative Forcing Effects on Climate (2003-2005) and also served as a member of the Committee on Earth Studies (1996-1999) and the Study on Transportation and a Sustainable Environment (1994-1997).

Dr. Daniel A. Jaffe is a professor of atmospheric and environmental chemistry at the University of Washington, Bothell. He is also an adjunct professor in the Department of Atmospheric Sciences at the University of Washington, Seattle. His areas of expertise are in global and regional atmospheric pollution, especially mercury; carbon monoxide; ozone; nitrogen oxides; aerosols and other metals; and long-range transport of air pollution in the Arctic and Pacific regions. Over the past 15 years he has been studying these pollutants at sites in Alaska, Russia, Japan, and several island stations in the Pacific Ocean. He received his Ph.D. in chemistry in 1987 from the University of Washington. His graduate work was concentrated in inorganic, analytical and atmospheric chemistry, atmospheric sciences, and environmental sciences and policy.

Dr. Sonia Kreidenweis is a professor of atmospheric science at Colorado State University. Her research focuses on characterization of the physical, chemical, and optical properties of atmospheric particulate matter, and the effects of the atmospheric aerosol on visibility and climate. She has conducted field studies in several U.S. national parks to establish the sources and characteristics of particulate matter responsible for visibility degradation, with a recent focus on the impacts of prescribed and wild fires. Ongoing laboratory and field studies have investigated the role of particles and of individual compounds found in particulate matter in the nucleation of cloud droplets and ice crystals. Prof. Kreidenweis is a past president of the American Association for Aerosol Research. She received her B.E. in chemical engineering from Manhattan College and her M.S. and Ph.D. in chemical engineering from the California Institute of Technology.

Dr. Katharine S. Law, moved to the University of Cambridge in 1987 to do a Ph.D. in atmospheric chemistry after studying environmental sciences and meteorology at university in the United Kingdom, Dr. Katharine S. Law worked for three years at the U.K. Meteorological Office. During her Ph.D. and subsequent years as a postdoc, senior research associate, NERC advanced research fellow at Cambridge, her research interests focused on quantifying the budgets and trends of trace gases such as tropospheric ozone and methane using numerical models and comparison with observations. She has taken part in the organization of many international airborne campaigns investigating processing in pollutant plumes. She has also been an active private investigator in the MOZAIC project since 1994. Since 2002 she has been employed as director of research by the Centre National de la Recherche Scientifique at Service d'Aeronomie in Paris. Here she has pursued research into the long-range transport of pollutants as part of the ICARTT/ITOP project in particular co-leading the Lagrangian field project and IGAC Task ITCT-2K4. She was also a co-coordinator for the participation of the M55-Geophysica aircraft during the AMMA campaign in summer 2006 (also an IGAC task) as well as leading the global chemical and aerosol modeling effort within AMMA-France/EU. Dr. Law is also co-chair of a new task POLARCAT-IPY that is focusing on transport of trace gases and aerosols to the Arctic in the framework of the International Polar Year.

Dr. Michael J. Prather is the Fred Kavli Chair and Professor in the Department of Earth System Science at University of California, Irvine. Dr. Prather has gained international recognition for research on atmospheric greenhouse gases, such as methane and ozone. As a member of the International Ozone Commission, Dr. Prather has participated in key U.N. environmental efforts. He has regularly addressed both government and business groups and is a scientific participant in major global environmental summits. Prior to joining the UC Irvine faculty Dr. Prather directed research at Harvard University and the Goddard Institute for Space Studies. A fellow of the American Geophysical Union and a member of the Norwegian Academy of Science and Letters, he served from 1997 through 2001 as editor-in-chief of *Geophysical Research Letters*. Dr. Prather has served on numerous NRC committees, including the Planning Group for the Workshop on Direct and Indirect Human Contributions to Terrestrial Greenhouse Gas Fluxes (Chair, 2003-2004), the Committee for Review of the U.S. Climate Change Science Program Strategic Plan (2002-2004), the Committee on EPA "Atmospheric Sciences" Workshop #1: Probabilistic Estimates of Climate Sensitivity (2002-2003), and the Board on Atmospheric Sciences and Climate (2000-2003).

Dr. Staci L. Simonich is an associate professor in the Department of Environmental and Molecular Toxicology and Chemistry at Oregon State University. Her current research focuses on understanding the atmospheric transport and deposition of semi volatile organic compounds to high-elevation ecosystems. The research program seeks to understand the relative impact of current and historical Asian, Pacific Ocean, and North American sources on contamination at high elevations. Dr. Simonich received her Ph.D. in chemistry from Indiana University in 1995. She received the Roy

F. Weston Environmental Chemistry Award from the Society of Environmental Toxicology and Chemistry in 2001 and the National Science Foundation Career Award in 2003.

Dr. Mark H. Thiemens is professor of chemistry and biochemistry and dean of the Division of Physical Sciences at the University of California, San Diego. He also directs UCSD's environmental science efforts in the interdisciplinary Center for Environmental Research and Training. As a scientist he is best known for his discovery of the mass-independent isotope effect, which has improved scientific understanding in areas as diverse as climate change, the origin of the solar system, chemical physics, acid rain, and the accumulation of greenhouse gases. The discovery led to his selection for the 1998 Ernest O. Lawrence Medal, the most prestigious award given to scientists by the U.S. Department of Energy. Dr. Tiemens received his Ph.D. in chemical oceanography from Florida State University, his M.Sc. from Old Dominion University, and his B.S. from the University of Miami. He was elected as a member of the National Academy of Sciences in 2006.

Staff

Dr. Laurie S. Geller is a senior program officer with the National Academies and serves within the Board on Atmospheric Sciences and Climate. She received a Ph.D. in analytical and atmospheric chemistry from the University of Colorado, Boulder in 1996. Her doctoral research (carried out with NOAA's Earth System Research Laboratory) focused on observations of the greenhouse gases nitrous oxide and sulfur hexafluoride. In 1996 she served as an AAAS Science Policy Fellow with the U.S. EPA's Office of Atmospheric Programs. From 2003 to 2008 she served as a science officer for the International Council for Science and taught at American University in Paris. She has directed several National Academies studies that addressed various aspects of atmospheric chemistry, climate change, and sustainable development.

Appendix E

Acronyms and Initialisms

ABLE	Atmospheric Boundary Layer Experiments
ACTO	Atmospheric Chemistry and Transport of Ozone
AIRS	Atmospheric Infrared Sounder
AQS	Air Quality Standards
ATOFMS	aerosol-time-of-flight mass spectrometry
CAA	Clean Air Act
CAIR	Clean Air Interstate Rule
CALIPSO	Cloud-Aerosol Lidar and Infrared Pathfinder Satellite Observations
CAMNET	Canadian Atmospheric Mercury Network
CAMR	Clean Air Mercury Rule
CARIBIC	Civil Aircraft for the Regular Investigation of the atmosphere Based on an Instrument Container
CASTNET	Clean Air Status and Trends Network
CCM	chemistry-climate model
CTM	chemical transport model
DI	dry air intrusion
EA	East Asia
EANET	Acid Deposition Monitoring Network in East Asia
EDGAR	Emissions Database for Global Atmospheric Research
EI	emissions inventory
STET	European Monitoring and Evaluation Programme

231

EOS	Earth Observing System
EPA	U.S. Environmental Protection Agency
ERS-2	European Remote Sensing satellite 2
ESRL	Earth System Research Laboratory
EU	European Union
EXPORT	European eXport of Precursors and Ozone by long-Range Transport

| FDDA | four-dimensional data assimilation |
| FOC | fluorinated organic chemical |

GAW	Global Atmosphere Watch
GEMS	Global and regional Earth-system (Atmosphere) Monitoring using Satellite and insitu data
GEOS	Goddard Earth Observing System
GHG	greenhouse gas
GOME	Global Ozone Monitoring Experiment

| HTAP-TF | Task Force on Hemispheric Transport of Air Pollution |

IADN	Integrated Atmospheric Deposition Network
IASI	Infrared Atmospheric Sounding Interferometer
ICARTT	International Consortium for Atmospheric Research on Transport and Transformation
IGAC	International Global Atmospheric Chemistry
IGACO	Integrated Global Atmospheric Chemistry Observations
IGOS	Integrated Global Observing Strategy
IMPROVE	Interagency Monitoring of Protected Visual Environments
INTEX-NA	Intercontinental Chemical Transport Experiment-NorthAmerica
IPCC	Intergovernmental Panel on Climate Change
ITCT	Intercontinental Transport and Chemical Transformation
ITCZ	Intertropical Convergence Zone
ITOP	Intercontinental Transport of Pollution

LDAR	Light Detection and Ranging
LRT	long-range transport
LRTAP	Long Range Transport of Air Pollutantion
LSM	Land surface model

| MAXOX | Maximum oxidation rates in the free troposphere |

MCS	mesoscale convective system
MDA8	maximum daily 8-hour average
MDN/AMI	Mercury Deposition Network/Atmospheric Mercury Initiative
METAALICUS	Mercury Experiment to Assess Atmospheric Loading in Canada and the U.S.
MLS	Microwave Limb Sounder
MODIS	Moderate Resolution Imaging Spectroradiometer
MOPITT	Measurements of Pollution in the Troposphere
MOZAIC	Measurements of OZone, water vapor, carbon monoxide and nitrogen oxides by in-service AIrbus airCraft
MOZART	Model for Ozone and Related Tracers
NA	North American
NAAQS	National Ambient Air Quality Standards
NARE	North Atlantic Regional Experiment
NARSTO	North American Research Strategy for Tropospheric Ozone
NASA	National Aeronautics and Space Administration
NEAQS	New England Air Quality Study
NEXAFS	Near-edge X-ray Absorption Spectroscopy
NH	Northern Hemisphere
NMHC	nonmethane hydrocarbon
NMVOC	nonmethane volatile organic compounds
NOAA	National Oceanic and Atmospheric Administration
NOXAR	Measurements of Nitrogen Oxides and Ozone Along Air Routes
NRC	National Research Council
NSF	National Science Foundation
OCP	organochlorine pesticide
OMI	Ozone Monitoring Instrument
PAH	polycyclic aromatic hydrocarbon
PBDE	polybrominated diphenyl ethers
PBL	planetary boundary layer
PCDD/Fs	polychlorinated dibenzo-p-dioxins and furans
PEM-West	Pacific Exploratory Mission-West
PFCA	perfluorinated carboxylic acid
PFOA	perfluorooctanoic acid
PHOBEA	Photochemical Ozone Budget of the Eastern North Pacific Atmosphere

PM	particulate matter
POLARCAT	Polar Study using Aircraft, Remote Sensing, Surface Measurements and Models of Climate Chemistry, Aerosols and Transport
POLINAT	Pollution from Aircraft Emissions in the North Atlantic Flight Corridor
POP	persistent organic pollutant
PRB	policy relevant background
PSD	Prevention of Significant Deterioration
RAINS	Regional Air Pollution Information and Simulation
RGM	Reactive Gaseous Mercury
RF	Radiative forcing
S-R	source-receptor
SA	South Asia
SCIAMACHY	Scanning Imaging Absorption Spectrometer for Atmospheric CHartographY
SH	Southern Hemisphere
SOAs	secondary organic aerosols
SONEX	subsonic assessment ozone and nitrogen oxide experiment
SRES	Special Report on Emissions Scenarios
STM	Scanning tunneling microscopy
TCO	tropospheric column ozone
TES	Tropospheric Emission Spectrometer
TMDL	Total Maximum Daily Load
TOMS	Total Ozone Mapping Spectrometer
TOPSE	Tropospheric Ozone Production about the Spring Equinox
TRACE-P	Transport and Chemical Evolution over the Pacific
TTN	Technology Transfer Network
UNEP	United Nations Environment Programme
UNECE	United Nations Economic Commission for Europe
UNFCCC	U.N. Framework Convention on Climate Change
VOC	volatile organic compound
WCB	warm conveyor belt
WHO	World Health Organization
WMO	World Meteorological Organization